Molecular Biology of Cancer

Mechanisms, Targets and Therapeutics

Lauren Pecorino

OXFORD
UNIVERSITY PRESS

OXFORD

UNIVERSITY PRESS

Great Clarendon Street, Oxford OX2 6DP

Oxford University Press is a department of the University of Oxford.
It furthers the University's objective of excellence in research, scholarship,
and education by publishing worldwide in

Oxford New York

Auckland Cape Town Dar es Salaam Hong Kong Karachi
Kuala Lumpur Madrid Melbourne Mexico City Nairobi
New Delhi Shanghai Taipei Toronto

With offices in

Argentina Austria Brazil Chile Czech Republic France Greece
Guatemala Hungary Italy Japan Poland Portugal Singapore
South Korea Switzerland Thailand Turkey Ukraine Vietnam

Oxford is a registered trademark of Oxford University Press
in the UK and in certain other countries

Published in the United States
by Oxford University Press Inc., New York

First published 2005
Reprinted 2006

British Library Cataloguing in Publication Data

Data available

Library of Congress Cataloging in Publication Data

Data available

Typeset by Laserwords Private Limited, Chennai, India

Printed in Great Britain
on acid-free paper by
Ashford Colour Press Limited, Gosport, Hampshire

ISBN 0-19-926472-4   978-0-19-926472-8

10 9 8 7 6 5 4 3 2

This book is dedicated to my mentors:

Raffaela and Joseph Pecorino

Professor Frank Erk

Professor Sidney Strickland

Professor Jeremy Brockes

In memory of:

Marie Favia

Mildred Maiello

Kerry O'Neill

■ OUTLINE CONTENTS

▪ DETAILED CONTENTS

■ PREFACE

The Molecular Biology of Cancer: Mechanisms, Targets and Therapeutics, is intended for both undergraduate and graduate level students (including medical students) interested in learning about how a normal cell becomes transformed into a cancer cell. Signaling pathways of a cell detect and respond to changes in the environment and regulate normal cellular activities. Cells contain many receptors on their membrane that allow a signal from outside of the cell (e.g. growth factors) to be transmitted to the inside of the cell. Signaling pathways are composed of molecules that interact with other molecules whereby one triggers the next in a sequence, in a way similar to the actions of team members of a relay race. The relay of information may cause a change in cell behavior or in gene expression and results in a cellular response (e.g. cell growth). Interference in these signal transduction pathways have grave consequences (e.g. unregulated cell growth) and may lead to the transformation of a normal cell into a cancer cell. The identification of the malfunctions of specific pathways involved in carcinogenesis provides scientists with molecular targets that can be used to generate new cancer therapeutics. I have chosen to present the biology of cancer together with a promise for its application towards designing new cancer drugs. Therefore, for most chapters in the text, the first half discusses the cell and molecular biology of a specific hallmark of cancer and the last half of the chapter discusses *therapeutic strategies*. To help form a link between particular molecular targets discussed in the first half of the chapter to the therapeutic strategies discussed in the last half of the chapter, a target symbol, ◎ is shown in the margin. I hope that this presentation stimulates interest and motivates learning of the subject matter.

Personally, I believe that the use of diagrams and illustrations is an extremely powerful tool of learning. A picture paints a thousand words...and more. I strongly suggest that the reader studies and enjoys the figures, artistically created by Joseph Pecorino. Major points and new cancer therapeutics are illustrated in red, and the target symbol, ◎ is used to identify molecular targets. Detailed descriptions of the figures are found in the body of the text.

Several additional features that are used throughout the text to facilitate learning and interest are described below:

Pause and think

These features are often presented in the margins of the text and are designed to engage the reader in thought and to present additional perspectives of core concepts. Many times questions are posed; sometimes they are answered and other times they encourage the re-reading of particular sections of text.

Special interest boxes

Shaded boxes are used to highlight special topics of interest such as the box entitled: *Skin Cancer* in Chapter 2. They are also used to provide additional explanation of more complex subjects such as *A little lesson about ROS...* in Chapter 2, *A little lesson about the MAP kinase family...* in Chapter 4, and *A little review of immunology basics...* in Chapter 10.

Lifestyle tips

These are suggestions about lifestyle choices and habits to minimize cancer risk, based on our current knowledge.

Leaders in the field of . . .

Scientists around the world have made contributions to the concepts presented in this text. Short biographies of several leading scientists, including their major contributions to a particular field of cancer biology, are presented. This feature is meant to give a human touch to the text. It may also be used as a tool for professional use and for following a continuing interest in the research literature. It may be of interest to listen to leading scientists in a particular subject area, by attending scientific conferences.

Analysis of . . .

Specific molecular techniques used to analyze particular biological and cellular events are described. It is important that science and medical professionals ask themselves "How do we know that?" Each of the major concepts underlying our current state of knowledge is the result of numerous experiments that generate data suggesting possible explanations and mechanisms of cellular events. The information retrieved is governed by the techniques that are used for analysis.

Chapter highlights: Refresh your memory

Summary points are listed in order to consolidate major concepts and provide a brief overview of the chapter. These may be particularly useful for revising for examinations.

Two features are included to strengthen your understanding of particular concepts presented in the chapter: Self tests presented within the text ask you to immediately reinforce material just presented and often refer to a figure. This causes a break from reading and engages you, the student, in "active" learning. An Activity that is aimed to strengthen your understanding of particular concepts and to encourage additional self-centered learning is presented towards the end of the chapter. Some require web-based research while others are more reflective.

Further reading is a list of general references found at the end of each chapter. These references consist mostly of reviews and support the contents of the chapter. They are not referenced in the main body of the text.

Selected special topics mainly lists specific primary research papers that *are* referenced in the main body of the text and may be pursued for further interest. Several relevant *web-sites* are also included.

Appendix 1 is a summary diagram that links key molecular pathways to the cell cycle.

Appendix 2 lists *Centers of Cancer Research* as a starting point for searching for research posts and employment in the field. Entries are separated by location (USA and UK).

Glossary

Over one hundred entries are defined in a clear and concise manner in order to provide students with a handy reference point for finding explanations of unfamiliar words.

It is my hope that the readers of this text will learn something new, become interested in something molecular, and ultimately, somehow, contribute to the field of cancer biology. The field of cancer biology evolves at a tremendous rate and so by the time of print, the information contained within these pages will need to be up-dated. This does not concern me because my aim is to present a *process* of how the pieces of science are put together and how we may attempt to apply our knowledge to cancer therapies. Many new drugs will fail but a select few will not. These select few will make marked improvements in the quality of life for many.

■ ACKNOWLEDGEMENTS

First, I would like to express my deepest gratitude to Jonathan Crowe, Commissioning Editor, at Oxford University Press. I am indebted to him for his faith that I could turn a one-page proposal into a complete textbook. He nurtured the synthesis of the book with special care, and provided a wealth of helpful suggestions and advice. With love, I thank my father, Joseph Pecorino, for his never-ending encouragement and I acknowledge his artistic talent used to translate dozens of my stick drawings into precise illustrations for the book during our visits across the Atlantic Ocean, over the last 2 years. Stephen Crumly kindly and faithfully reproduced the illustrations using his fine skills in computer graphics under tight deadlines. Thanks also to Sally Lane, Production Editor, and her production staff and Jonathan Rubery and Ruth Trigg, for additional assistance, at Oxford University Press.

Kind appreciation is expressed for the precise and critical comments given by my official reviewers: Tony Bradshaw, Oxford Brookes University, UK, Moira Galway, St. Francis Xavier University, Canada, Maria Jackson, University of Glasgow, UK, Helen James, University of East Anglia, UK, and Ian Judson, Cancer Research UK, London, UK. The value added to the text by these scientists cannot be underestimated. Their comments have had tremendous impact.

MaryBeth Maiello provided me with motivation and scientific advice from the very start of this project. I am truly fortunate to have met Ken Douglas, University of Manchester, whose willingness to read and offer comments on many chapters was astonishing. I thank colleagues at the University of Greenwich, especially Mark Edwards, Laurence Harbige, and Mike Leach for many helpful suggestions, and also John Nicholson and Richard Blackburn for their support. Students Sarah Cheeseman, Sarah Thurston, and Dario Tuccinardi also provided valuable feedback on selected chapters.

Many fellow scientists have made suggestions or other contributions to the book, including: Tim Crook, Jeremy Griggs, several members of the Kuriyan laboratory, Gerd Pfeifer, Mariann Rand-Weaver, and Jerry Shay. Special thanks are expressed to J. Griggs and M. Rand-Weaver for helping me to obtain electronic figures. The Wellcome Library, London, provided an ideal scientific sanctuary. I admire and acknowledge the work of all those scientists whose research efforts have contributed to the field of cancer research.

I am especially grateful to the support that came from my family, especially Teresa Rapillo, and from friends. Marcus Gibson, a truly superlative man, made the completion of this book possible and his support has been invaluable.

1

Introduction

Introduction

The aim of this text is to provide a foundation in the molecular biology of cancer and to demonstrate the conceptual process that is being pursued in order to design more specific cancer drugs. Common threads are woven throughout the different chapters so that the terminology becomes familiar and the logic of cellular mechanisms becomes clear. The text also provides guidance in everyday decisions that may lead to a decrease in cancer risk. The translation of the knowledge of molecular pathways into clinically important therapies (linked throughout the text by the target symbol, "◎") will be communicated and will breathe excitement into learning. Academically, you will gain a foundation in the cell and molecular biology of cancer. More importantly, you will develop an intellectual framework upon which you can add new discoveries that will interest you throughout your lifetime. My goal in writing this book is to inspire. It would be most gratifying for me if, by reading this book, you the reader would be compelled to contribute to the cancer research field directly. Knowledge is powerful.

Cancer statistics are shocking. Cancer affects one in three people. It is estimated that 555,500 Americans died from cancer in 2002 and the mortality rates (number of cancer deaths per year per 100,000 people) were over 200 for the UK in the 1990s. The worldwide incidence (number of new cases) is about 10 million cases per year, a figure which is due to double in 20 years. These numbers are cold, stark and impersonal. Hidden behind them are tears, fears, pain and loss. No one is excluded from the risk. There is a need to understand the disease and to translate our knowledge into effective therapies. In order to understand the process of carcinogenesis, whereby a normal cell is transformed into a cancer cell, we must know the intricacies of cell function and the molecular pathways that underlie it. We must consider the cell in the context of the entire body. We have a lot to learn! However, the knowledge of the molecular details in important cellular and biochemical pathways can be applied to a new wave of cancer therapies. What better reward for these efforts?

1.1 What is cancer?

Cancer is a group of diseases characterized by unregulated cell growth and the invasion and spread of cells from the site of origin, or primary site, to other sites in the body. Several points within this definition need to be emphasized. First, cancer is considered a group of diseases. Over 100 types of cancer have been classified. The tissue of origin gives the cancer distinguishing characteristics. Approximately 85% of cancers occur in epithelial cells and are classified as carcinomas. Cancers derived from mesoderm cells (e.g. bone, muscle) are called sarcomas and cancers of glandular tissue (e.g. breast) are called adenocarcinomas. Cancers of different origins have distinct features. For example, skin cancer has many characteristics that differ from lung cancer. The major factor that causes cancer in each target tissue is different: ultraviolet (UV) radiation from the sun can easily target the skin while inhalation of cigarette smoke can target the lungs. In addition, as will be examined in detail later, there are differences in the molecular mechanisms involved in carcinogenesis within each cell type and the pattern of cell spread from the primary site. Treatment must be applied differently. Surgical removal of a cancerous growth is more amenable to the skin than the lungs. This initial view presents layers of complexity which may seem insurmountable to dissect in order to improve the conventional therapeutic approaches. However, even though the underlying cellular and molecular routes are different, the end result is the same. Upon fine analysis, Hanahan and Weinberg (2000) have defined six hallmarks of most, if not all, cancers (Figure 1.1). They propose that

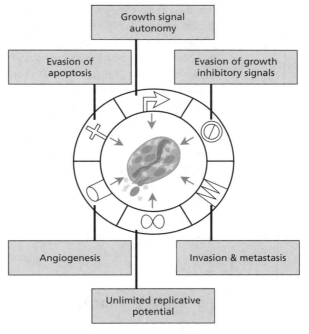

Figure 1.1 The six hallmarks of cancer. Reprinted from *Cell*, Vol. 100, Hanahan *et al.*, "The Hallmarks of Cancer", p. 57, Copyright (2000), with permission from Elsevier Science

acquiring the capability for growth signal autonomy, evasion of growth inhibitory signals, evasion of apoptotic cell death, unlimited replicative potential, angiogenesis (the formation of new blood vessels), and invasion and metastasis are essential for carcinogenesis. Each will be examined in detail in this text and each is a potential target pathway for the design of new therapeutics.

The six hallmarks of cancer (see Figure 1.1)

* Growth signal autonomy
    - Normal cells need external signals from growth factors to divide
    - Cancer cells are not dependent on normal growth factor signaling
    - Acquired mutations short circuit growth factor pathways leading to unregulated growth

* Evasion of growth inhibitory signals
    - Normal cells respond to inhibitory signals to maintain homeostasis (most cells of the body are not actively dividing)
    - Cancer cells do not respond to growth inhibitory signals
    - Acquired mutations interfere with the inhibitory pathways

* Evasion of apoptosis (programmed cell death)
    - Normal cells are removed by apoptosis, often in response to DNA damage
    - Cancer cells evade apoptotic signals
    - Loss of apoptotic regulators through mutation lead to evasion of apoptosis

* Unlimited replicative potential
    - Normal cells have an autonomous counting device to define a finite number of cell doublings after which they become senescent. This cellular counting device is the shortening of chromosomal ends, telomeres, that occurs during every round of DNA replication
    - Cancer cells maintain the length of telomeres
    - Altered regulation of telomere maintenance results in unlimited replicative potential

* Angiogenesis (formation of new blood vessels)
    - Normal cells depend on blood vessels to supply oxygen and nutrients but the vascular architecture is more or less constant in the adult
    - Cancer cells induce angiogenesis, the growth of new blood vessels, needed for tumor survival and expansion
    - Altering the balance between angiogenic inducers and inhibitors can activate the angiogenic switch

* Invasion and metastasis
    - Normal cells maintain their location in the body and generally do not migrate →

- → The movement of cancer cells to other parts of the body is a major cause of cancer deaths
- Mutations alter the activity of enzymes involved in invasion and alter molecules involved in cell–cell and cell–extracellular adhesion

PAUSE AND THINK

Why are malignant tumors life-threatening? They are physical obstructions and as they invade other organs they compromise function. They also fiercely compete with healthy tissues for nutrients and oxygen.

Cancer is characterized by unregulated cell growth and the invasion and spread of cells from the site of origin. This leads to the distinction between a benign tumor and a malignant tumor. A benign tumor is not evidence of cancer. Benign tumors do not metastasize, although some can also be life-threatening. Malignant tumors, on the other hand, do not remain encapsulated, show features of invasion and metastasize.

Cancer cells can be distinguished from normal cells in cell culture conditions

Normally, cells grow as a single layer, or monolayer, in a petri dish due to a property called contact inhibition; contact with neighboring cells inhibit growth.

Transformed cells acquire the following phenotypes:

- they fail to exhibit contact inhibition and instead grow as piles of cells or "foci" against a monolayer of normal cells
- they can grow in conditions of low serum
- they adopt a round morphology rather than flat and extended
- they are able to grow without attaching to a substrate (e.g. the surface of a petri dish), exhibiting anchorage independence

1.2 Evidence suggests that cancer is a genetic disease at the cellular level

Interestingly, most agents that cause cancer (carcinogens) are agents that cause alterations to the DNA sequence or mutations (mutagens). Thus, similarly to all genetic diseases, cancer results from alterations in DNA. A large amount of evidence indicates that the DNA of tumor cells contains many alterations ranging from subtle point mutations (changes in a single base pair) to large chromosomal aberrations such as deletions and chromosomal translocations. The accumulation of mutations in cells over time represents a multi-step process that underlies carcinogenesis. The requirement for an accumulation of mutations explains why there is an increased risk of cancer with age and why cancer has become more prevalent over the centuries as human lifespan has increased. There have been more cases of cancer in recent years because we are living longer. The longer we live the more time there is to expose our DNA to accumulating mutations which may lead to cancer. Interestingly, only 5–10% of the mutations observed are thought to be directly involved in causing cancer based upon mathematical modeling. This estimate provides the basis for the current optimism in the field of molecular therapies. Almost all of the mutations identified in tumor cells are somatic mutations whereby the DNA of a somatic (body) cell has been damaged. These mutations are not passed on to the next generation of offspring, and therefore are not inheritable, but they are passed to daughter cells after cell division. Thus, cancer is considered to be a genetic disease at the cellular level. Only alterations in the DNA of sperm or egg cells, called germ-line mutations, will be passed on to offspring. Some germ-line mutations can cause an increased risk of developing cancer but are rarely involved directly. Cancer cells continue to change their behavior as they progress. The progressive changes of a cell resulting from an accumulation of genetic mutations that confer a growth advantage over its neighbors proceeds in a fashion analogous to Darwinian evolution: chance events give rise to mutations that confer changes in phenotype and allow adaptation to the environment, resulting in the selection and survival of the fittest. This classifies the mechanism of cancer as obeying "natural order" and being statistically inevitable and is discussed at length by Mel Greaves, in his book *Cancer, the evolutionary legacy*. The accumulation of mutations occurs only after the cell's defense mechanisms (e.g. DNA repair) are evaded. Any alterations of DNA that are not repaired before the next cell division are passed on to the daughter cells and are perpetuated. The cell relies on several processes to repair damaged DNA. In cases of severe DNA damage, cell suicide is induced in order to protect the whole body from cell transformation. The molecular details of these processes and the mutations that compromise them will be described in Chapters 2 and 6. Thus, many mechanisms exist for blocking carcinogenic events but over-burdening the system increases the probability that a cell carrying a deleterious mutation will escape surveillance.

Growth, apoptosis, and differentiation regulate cell numbers

There are three important processes that contribute to the overall net cell number in an individual. Cell proliferation (cell division, cell growth) is the most obvious. Cell division results in two daughter cells. Secondly, the elimination of cells by programmed cell death also affects the net cell number. Lastly, during the process of differentiation cells can enter an inactive phase of cell growth and thus differentiation can affect net cell numbers. DNA mutations that alter the function of normal genes involved in growth, apoptosis, or differentiation can affect the balance

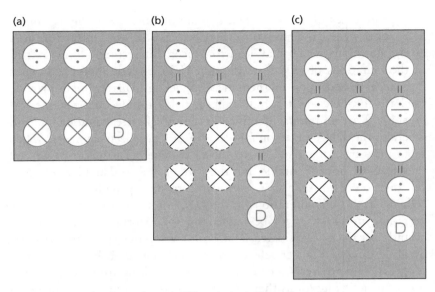

Figure 1.2 Growth, apoptosis, and differentiation affect cell number

of cell numbers in the body and lead to unregulated growth. Examine the simplistic model above (Figure 1.2). If four of the nine cells shown in Figure 1.2a divide ⊕, and four are programmed to die by apoptosis ⊗, and one differentiates ⒟ (so the cell neither dies nor divides) the cell number will remain the same (Figure 1.2b; remaining cells shown in red). However, if apoptosis is blocked in one cell and that cell divides instead, the total number of cells will increase to 11 (Figure 1.2c). Similarly if differentiation is blocked in a cell and that cell divides, as is the case of some leukemias, the cell number will also increase. Thus an alteration in the processes of growth, apoptosis, or differentiation, can alter cell numbers. Normal genes that can be activated by mutation to be oncogenic are called proto-oncogenes. Proto-oncogenes play functional roles in normal cells. The term reminds us that all normal cells have genes that have the potential to become oncogenic.

The sequence of stages through which a cell passes between one cell division and the next is called the cell cycle (Figure 1.3) and is made up of four stages: G1, S phase, G2, and M phase. M phase includes mitosis and cytokinesis. G1, S, and G2 make up the rest of the cycle called interphase. In Figure 1.3, the length of the cell cycle is 16 h (1 h for M phase and 15 h for interphase), but note that this can vary depending on the cell type. Most cells in the adult are not in the process of cell division. They are quiescent and enter an inactive period called G0. Mitogens or growth factors can, however, induce the cells to re-enter the cell cycle and pass a control point called the G1 restriction point. The passage of the cell through the different phases of the cell cycle is coordinated and

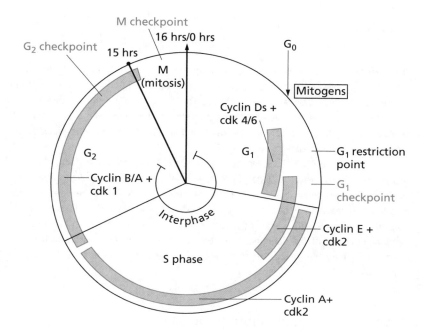

Figure 1.3 The cell cycle

regulated by a set of proteins called cyclins and their associated cyclin-dependent kinases (cdks). The concentrations of the cyclins rise and fall in a regular pattern during the progression of the cell cycle, hence their name. The pairing of cyclins to the cyclin-dependent kinases is highly specific. The different cyclin-cdk complexes drive the cell through different phases of the cell cycle as shown in Figure 1.3. Note that cyclin D is the first cyclin to be synthesized and functions in the G1-S phase transition. Cyclin D plays a role in the regulation of expression of the cyclin E gene. Interestingly, overexpression of cyclin D has been reported in some cancers.

Cell cycle checkpoints (see Figure 1.3), a series of biochemical signaling pathways that sense and induce a cellular response to DNA damage, are important for maintaining the integrity of the genome. The G1 checkpoint leads to the arrest of the cell cycle in response to DNA damage ensuring DNA damage is not replicated during S phase. The G2 checkpoint leads to the arrest of the cell cycle in response to damaged and/or unreplicated DNA to ensure proper completion of S phase. The M checkpoint leads to the arrest of chromosomal segregation in response to misalignment on the mitotic spindle. The components of the checkpoints are proteins that act as DNA damage sensors, signal transducers, or effectors. Disruption of checkpoint function leads to mutations that can induce carcinogenesis.

Oncogenes and tumor suppressor genes

Growth is regulated by both positive and negative molecular factors. Thus, to increase growth, enhancement of positive factors or depletion of negative factors are required. There are two major classifications of mutated genes that contribute to carcinogenesis: oncogenes and tumor suppressor genes (Figure 1.4). A general description of an oncogene is a gene mutated such that its protein product is produced in higher quantities, or has increased activity and therefore acts in a dominant manner to initiate tumor formation. For example, one oncogene produces increased

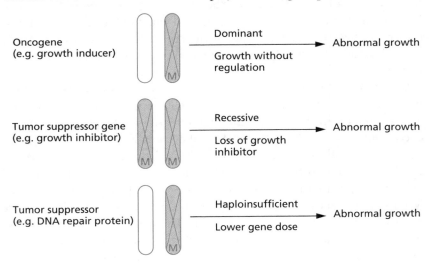

Figure 1.4 Oncogenes and tumor suppressor genes

quantities of a specific growth factor (e.g. platelet-derived growth factor) which stimulates growth inappropriately. Another example is an oncogene that produces a growth factor receptor with increased activity because it has been altered so that it is always in the "on" state and does not require growth factor to transduce a signal into the cell. "Dominant" refers to the characteristic that a mutation in only one allele is sufficient for an effect.

Analysis of oncogene function by *in vitro* cell transformation assays

The prototypical experiment used to demonstrate the presence of an oncogene is to test for cell transformation in culture. DNA of interest is isolated and introduced into a standard cell line called NIH/3T3 cells (mouse fibroblast cells) by calcium phosphate precipitation or electroporation. If the test DNA contains an oncogene, foci (mentioned above) will form and will be easily identifiable against a monolayer of untransformed NIH/3T3 cells.

Tumor suppressor genes code for proteins that play a role in inhibiting both growth and tumor formation. Loss of growth inhibition occurs when mutations cause a loss of function of these genes. Consequently, growth is permitted. Tumor suppressor mutations are mainly recessive in nature because one intact allele is usually sufficient to inhibit growth, thus both alleles of the gene must be mutated before the loss of function is actually seen phenotypically. Recessive mutations support Knudson's two-hit hypothesis, the classical model used to explain the mechanism behind tumor suppressor action (see Chapter 5 and Figure 5.2). It states that both alleles need to be mutated (recessive) to trigger carcinogenesis. This model has been used to explain the mechanism behind conditions that predispose individuals to increased cancer risk. Patients inherit one mutated tumor suppressor allele and may acquire a second somatic mutation over time. Therefore, these patients have a "head start" towards a cancer phenotype in the race of mutation accumulation. Recent evidence suggests there is an alternative mechanism for particular tumor suppressor genes called haploinsufficiency, whereby only one mutated allele can lead to the cancer phenotype. As the term suggests, one normal allele produces half (haplo) of the quantity of protein product produced by normal cells and this is not enough to suppress tumor formation in these cases. This has been demonstrated for genes that regulate DNA repair and the DNA-damage response such that reduced activity leads to genetic instability. Gene dosage may also affect the spectrum of tumors observed; haploinsufficiency may cause cancer in some cell types and recessive mutations may cause cancer in other cell types (Fodde and Smits, 2002).

PAUSE AND THINK

Analogies can be used to illustrate the mechanisms of oncogenes and tumor suppressor genes. Traffic, like cell growth, is also regulated by "stop" and "go" signals. A single traffic light gets stuck on green and cars continue to travel through an intersection unregulated. This is analogous to the dominant mechanism of an oncogene that signals unregulated growth. Imagine there are two inhibitory traffic signals on a road. Removal of both signals are necessary before a driver would proceed (removal of one leaves one "stop" signal remaining). This is analogous to the recessive mechanism of a tumor suppressor gene, whereby mutation of both alleles results in the release of growth inhibition.

All of the cancer cells in a patient arise from a single cell that contains an accumulation of initiating mutations; in other words, the development of cancer is clonal. It is generally assumed that only one of the 10^{14} cells in the body needs to be transformed in order to create a tumor. However, studies of adult stem cells have made recent contributions to our understanding of carcinogenesis. Stem cells are undifferentiated cells that have the ability to self-renew and produce differentiated progeny. Normal stem cells may be a main starting point for carcinogenesis in some cancers since both cancer cells and stem cells utilize and rely on self-renewal molecular programs. Also, cancer is more likely to develop in cells that are actively proliferating since there is a greater chance for mutations to accumulate; normal stem cells continue to proliferate over long periods of time. These concepts will be discussed further in Chapter 7.

The concepts described above suggest that cancer is a genetic disease at the cellular level.

1.3 Influential factors in human carcinogenesis

Environment, reproductive life, diet, and smoking are four factors that play an important role in carcinogenesis. These lifestyle factors, in principle, can be altered to prevent most cancers. Exposure to carcinogens, hormonal modifications influenced by childbirth and birth control, and exposure to viruses, underlie these lifestyle factors. Epidemiology, the study of disease in the population, has been instrumental in elucidating the contributions of these factors towards different cancers. Although molecular details will be discussed in later chapters, a brief introduction of each factor is given below.

Environment

Observations by a British surgeon in 1775 resulted in the first correlation between an environmental agent and specific cancers. Percival Pott concluded that the high incidence of nasal and scrotum cancer in chimney sweeps was due to chronic exposure to soot. Not only where you work, but also the choice of where you relax can contribute to your risk of cancer. Unprotected exposure to the sun exposes your skin to UVB radiation which can directly alter your DNA by forming pyrimidine dimers and cause mutations. Sun blocks that have UV absorbing ingredients have been developed to protect your skin from UV radiation and are a good defense if you do decide to relax in the sun.

Reproductive life

Another early observation was that nuns are more likely to develop breast cancer than other women. We now know that having children reduces breast cancer risk for women compared with not having children. The age of a woman at the time of giving birth for the first time and the age of a woman at the initiation and termination of her menstrual cycles also influences cancer risk. Hormonal contraception and fertility treatments also affect cancer risk because they alter a women's ovulation schedule (active ingredients prevent and promote ovulation, respectively). Sexual promiscuity can also contribute to increased risk of cancer. Sexually transmitted human papillomaviruses can be found in all cervical cancers worldwide. It is not surprising therefore that nuns have a low incidence of cervical cancer. Barrier methods of contraception can protect against this infectious pathogen.

Diet

The incidence of a specific cancer varies greatly between different populations in different geographical locations. Observation of immigration patterns has revealed that local cancer rates strongly influence cancer risk with diet being one of the most influential factors. Figure 1.5 shows a comparison of cancer prevalence between US females and Chinese males (King, 2000). Stomach cancer is the predominant cancer in the Chinese population and a minor cancer in the population of the USA. Interestingly, the risk of stomach cancer in Chinese people who have migrated to the USA decreases only if they adopt the American diet, but not if they retain an Eastern diet. The Mediterranean diet which is rich in fresh fruit, vegetables and red wine has been promoted to be beneficial in reducing cancer risks. Recently, studies of the molecular interactions of individual dietary constituents (e.g. polyphenols, carotenoids, and allium compounds) with cellular signaling pathways have begun and will be examined in Chapter 9.

Smoking

The clearest example of lifestyle factors underlying a specific cancer is the discovery that smoking causes lung cancer (it is also implicated in

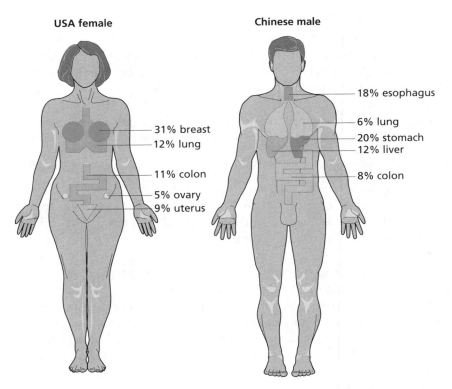

USA female   Chinese male

31% breast
12% lung
11% colon
5% ovary
9% uterus

18% esophagus
6% lung
20% stomach
12% liver
8% colon

Figure 1.5 Predominant cancers differ among different populations. Data from King, 2000

pancreas, bladder, kidney, mouth, stomach, and liver cancer). Smoking accounts for 40% of all cancer deaths. At least 81 carcinogens have been identified in cigarette smoke. Smoking became particularly fashionable in Europe and the USA during World War I and World War II and resulted in an epidemic of lung carcinoma. After vast public education campaigns and a subsequent reduction of smoking, lung cancer rates have fallen dramatically. It is clear that some future cancer deaths can be avoided by changes in lifestyle factors.

Additional influences

In addition to lifestyle factors, there are risk factors inherent in our own physiology. By-products of our metabolism and errors that occur during DNA replication contribute to carcinogenesis. Aerobic metabolism produces oxygen radical by-products that are mutagenic. Several inherited metabolic diseases also produce mutagenic by-products. For example, tyrosinemia type I patients have a defect in the enzyme fumarylacetoacetate hydrolase which is involved in tyrosine breakdown. As a result of this block, the by-products fumarylacetoacetate and maleylacetate accumulate and increase cancer risk due to their ability covalently to modify DNA and cause mutations. During DNA replication and repair, polymerases can introduce mutations directly in DNA because of their associated error rates. There is a perpetual inherent risk of mutation during the lifespan of a cell by the nature of cell processes.

1.4 Principles of conventional cancer therapies

The earliest therapeutic strategy used against cancer was surgically to remove as much of the cancer as possible. Obviously this is easier in some types of cancer and impossible in other types. It is not a precise procedure at the cellular level and does not address the question of cells that have spread from the primary site (metastasized cells). Therefore chemotherapy and radiotherapy have been used to inhibit or eradicate metastasized cells. The objectives of cancer therapies are to prevent proliferation (cytostatic effect) and to kill the cancer cells (cytotoxic effect). The aim for all drugs is to achieve an effective result with minimum side effects. This is indicated by the therapeutic index. It is the value of the difference between the minimum effective dose and the maximum tolerated dose (MTD) (Figure 1.6). The larger the value, the better the drug. Many conventional cancer treatments are administered at maximum tolerated doses (MTDs).

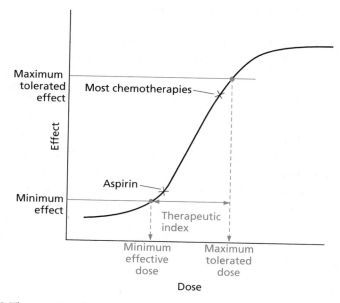

Figure 1.6 Therapeutic index

PAUSE AND THINK

Let us "create" an example purely to illustrate the concept of the MTD. Two aspirins may be the minimum for an effective dose against a headache and 30 may be the dose that can be tolerated before harmful side effects are observed. However, if harmful side effects were seen after three aspirins the therapeutic index would decrease and the drug would be much less favorable.

Chemotherapy

Conventional chemotherapy uses chemicals that target DNA, RNA, and protein to disrupt the cell cycle in rapidly dividing cancer cells and thus has broad specificity. The ultimate goal of cytotoxic chemotherapy is to cause severe DNA damage and to trigger apoptosis in the rapidly dividing cancer cells. The side effects of chemotherapy, which we are all too aware of, such as alopecia (loss of hair), ulcers, and anemia, are due to the fact that hair follicles, stomach epithelia, and haemopoietic cells are also rapidly dividing and therefore they too are greatly affected by these drugs. Often prescribed MTDs induce toxicity in sensitive tissues and require a pause in drug administration so that normal cells can recover. We must appreciate that conventional chemotherapies (e.g. cisplatin, methotrexate; discussed in Chapter 2) have had results in treating cancers and continue to extend lives but at the same time we must desire drugs with better efficiencies and less severe and debilitating side effects. We must strive to develop drugs that rise to these expectations.

1.5 The role of molecular targets in cancer therapies

The major flaw in the rationale of most conventional therapies is the focus on a conceptional race between normal cells and transformed cells: can we kill the transformed cells quicker than normal cells? As a result the side effects of most therapies are very harsh, as mentioned above. There is a need to learn about the differences between normal and transformed cells at the molecular level in order to identify cancer-specific molecular targets. In this way, we can design drugs that will be specific for the cancer cells, have increased efficacy and cause fewer side effects.

Clinical trials

Testing of new drugs must progress through staged clinical trials (Table 1.1). Phase I trials examine dose responses for assessing drug safety, using a small number (20–80) of healthy volunteers or patients. Many parameters of the metabolism of the drug in humans (e.g. How long does the drug remain in the body?) are obtained at this time. About 70% of drugs tested in Phase I will progress to Phase II studies. Phase II trials are designed to examine efficacy in a larger group of people (100–300). Phase III trials should not be initiated prior to knowing the effective drug dosage. Phase III trials are large-scale studies (1000–3000 people) to confirm drug effectiveness, monitor side effects, and also compare the efficacy of the new drug to conventional

Table 1.1 Clinical trials

	Purpose	Number of patients
Phase I	Safety	20–100
Phase II	Efficacy	Up to several hundred
Phase III	Efficacy often tested against conventional treatments	Several hundred to several thousands

treatments. Only terminally ill patients may be recruited for clinical trials, by law, in many countries. This has implications for the outcome of testing particular drugs and will be discussed later in the text. About 30% of drugs tested successfully complete Phase III studies. Drugs are also tested against control populations. These people either receive no treatment or receive a placebo, or inactive substance. In order to reduce the risk of bias, trials can be randomized. That is, patients are randomly assigned to either a treatment or control group, guaranteeing that the two groups are similar. In addition, the trial may be conducted as a single-blind study, whereby patients do not know which group they are in, or as a double-blind study, whereby neither patients nor investigators know who has received the treatment or placebo until after a code is broken that identifies the people in the two groups. Unlike conventional chemotherapies, targeted therapy may not require maximum tolerated doses. Experience is teaching us that the design of the trials for molecularly targeted drugs needs to be well thought out. It is important to consider the stage and type of cancer to be treated, patient populations with the correct molecular profiles (that is, does the patient carry mutations in genes of interest), assessment of the compound in inhibiting its molecular target, and assessment of the relationship between molecular inhibition and clinical response.

Molecules of fame

As we examine the molecular pathways that underlie carcinogenesis we must keep in mind that the pathways do not act in isolation but are interconnected (see Appendix 1). Despite the many hundreds of molecules involved in carcinogenesis, there are several families of "star players" in the story of carcinogenesis. Many of these "star players" act as nodes that receive signals from many pathways and can exert several effects in response to a specific signal.

PAUSE AND THINK

Is tumor shrinkage a suitable assessment criteria for cytostatic drugs? No, because direct cell death is not anticipated by this class of drugs. Cytotoxic drugs are expected to kill tumor cells.

The family of protein kinases, one of the largest families of genes in eukaryotes, must be included in any introduction to cancer biology. Protein kinases phosphorylate (add a phosphate group to) a hydroxyl group on specific amino acids in proteins. Tyrosine kinases phosphorylate tyrosine residues while serine/threonine kinases phosphorylate serine and threonine residues. Phosphorylation results in a conformational change and is an important mechanism for regulating the activity of a protein. Kinases can be found at the cell surface as transmembrane receptors, inside the cell as intracellular transducers, or inside the nucleus. Kinases play a critical role in major cell functions including cell cycle progression, signal transduction, and transcription, and are important molecular targets for cancer drug design. Phosphorylation is also regulated by phosphatases, enzymes that remove phosphate groups. Mutational analysis of all known human protein tyrosine phosphatases suggests that several of these act as tumor suppressors in some types of cancer (Wang *et al.*, 2004).

The Ras family is another "star" set of genes which are found mutated in over 50% of certain cancers (e.g. colon cancer). Ras is an intracellular transducer protein that acts subsequently to binding of a growth factor to its receptor and is involved in transmitting the signal from the receptor through the cell. As G proteins, they reside on the intracellular side of the plasma membrane and are activated by the exchange of GDP for GTP.

Tumor Protein p53 (TP53; p53) and its related family members hold pivotal positions in guarding the integrity of the genome by coordinating responses of the cell (e.g. cell cycle arrest, DNA repair, apoptosis) to different types of stress (e.g. DNA damage, hypoxia). The p53 gene is a tumor suppressor gene that has a key role in inhibiting carcinogenesis. It is mutated in more than half of all cancers and over a thousand different mutations have been identified. It acts as a transcription factor and induces the expression of genes required to carry out its functions.

The retinoblastoma gene is also a tumor suppressor gene that plays a central role in regulating the cell cycle. It is commonly mutated in several cancers. The retinoblastoma protein normally functions as an inhibitor of cell proliferation by binding to and suppressing an essential transcription factor of cell cycle progression. Its activity is regulated by phosphorylation by cyclin D and the cyclin-dependent kinases (4/6).

Introduction of cancer genomics

The completion of the Human Genome Project, whereby every nucleotide of a human genome has been sequenced and mapped, has paved the way for cancer genomics. Learning about the details of the genome of a cancer cell and how it differs from a normal cell will provide us with the fine distinctions needed to design more powerful and specific drugs. Several projects are underway. The Cancer Genome Anatomy Project (CGAP) is aimed at developing cancer-specific gene data sets for public access (Strausberg *et al.*, 2001). Both academia and biopharmaceutical industries can mine the databases and utilize the online informatics tools. It is envisaged that *in silico* (computer) analysis of these data sets will help define specific molecular signatures for specific cancers and promote the development of new methods of diagnostics and treatments. New human genes and tumor-specific gene deregulation have been discovered through use of the CGAP database. The human genome has recently been subdivided into the "kinome" by the mapping of the complete set of 518 protein kinase genes in the genome (Manning *et al.*, 2002). Since aberrant regulation and mutation of these genes are involved in carcinogenesis, this will be an important tool for the design of new molecular therapies. The International SNP Map Working Group is analyzing single nucleotide polymorphisms (SNPs) to identify mutations within the genome that may be linked with cancer. Recent discoveries on the functional role of small RNAs have led to new tools (e.g. RNA interference) to help elucidate the function of genes in an organism. The findings from studies of functional genomics promises to provide insights into cancer biology.

Analysis of gene function by small interfering RNAs (siRNAs)

Gene function is often determined by abolishing the expression of a gene product and observing the resulting phenotype. RNA interference is a cellular mechanism for regulating gene expression in most eukaryotes. Short RNA duplexes (approximately 21 nucleotides long with two nucleotide 3' overhangs) called small interfering RNAs mediate the expression of genes by causing the degradation of homologous single-stranded target RNAs. Experimentally, we can use siRNAs to target endogenous genes in mammalian cells. Using the known sequence of a segment of target mRNA, sense and antisense RNAs are designed, synthesized, and annealed to produce siRNA duplexes. The siRNAs are delivered to cells by classical gene transfer methods (e.g. electroporation). Specific antibodies against the targeted protein are often used to ensure that target proteins levels have been diminished. See Kamath *et al.* (2003) as an example of the use of this approach.

The nature of the next generation of drugs is likely to be small molecules targeted against selective gene products and/or is likely to utilize methods (e.g. antisense RNA) to prevent expression of target genes. Several therapies of this design that can treat specific cancers already exist and will be described later in the text.

■ **CHAPTER HIGHLIGHTS—REFRESH YOUR MEMORY**

- Cancer is a common disease that will affect one out of three people worldwide.

- Cancer is a group of diseases that results in the spreading of mutated cells throughout the body.

- There are six hallmarks of cancer. They are:

  - Growth signal autonomy

  - Evasion of growth inhibitory signals

  - Evasion of apoptosis

  - Unlimited replicative potential

  - Angiogenesis

  - Invasion and metastasis.

- Most carcinogens are mutagens.

- Carcinogenesis is a multi-step process that requires the accumulation of several mutations.

- Cancer is a genetic disease at the cellular level.

- Genes that are involved in growth, differentiation or cell death when deregulated can give rise to the cancer phenotype.

- The progression of a cell through the different phases of the cell cycle is highly regulated by cyclins and cyclin-dependent kinases.

- A gene containing a dominant mutation that results in inappropriate activation of growth is an oncogene.

- Tumor suppressor genes are usually inactivated by mutations in two both alleles (recessive) and this results in inactivation of growth inhibition.

- Haploinsufficiency, whereby only one allele of a tumor suppressor gene is inactivated, also contributes to carcinogenesis.

- Changes in lifestyle factors can affect cancer risk.

- Many conventional therapies are broad-acting drugs administered at MTD resulting in severe side effects.

- Protein kinases, enzymes that phosphorylate proteins, are important molecules in carcinogenesis.

- Cancer genomics is being used to define molecular targets for tumor specific effects.

- Some cancers can already be treated by specific molecular approaches.

■ **ACTIVITY**

Become familiar with the Cancer Genome Anatomy Project web site.
On which chromosome is the p53 gene located?
Is it expressed in bone and mammary gland tissues? Name three genes regulated by this transcription factor.

■ **FURTHER READING**

Alison, M.R. (2002) *The Cancer Handbook*, Nature Publishing Group, Macmillan Publishers, London.

Hanahan, D. and Weinberg, R.A. (2000) The hallmarks of cancer. *Cell* **100**:57–70.

King, R.J.B. (2000) *Cancer Biology*, 2nd edn. Pearson Education Ltd, London.

Peto, J. (2001) Cancer epidemiology in the last century and the next decade. *Nature* **411**:390–395.

Reddy, A. and Kaelin W.G., Jr. (2002). Using cancer genetics to guide the selection of anticancer drug targets. *Curr. Opin. Pharm.* **2**:366–373.

■ WEB SITES

The Cancer Genome Anatomy Project web site homepage http://cgap.nci.nih.gov
The Food and Drug Administration web site http://www.fda.gov/oc/gcp/education.html
American Cancer Society; Cancer Statistics 2004
http://www.cancer.org/docroot/pro/content/pro_1_1_Cancer_Statistics_2004_presentation

■ SELECTED SPECIAL TOPICS

Kamath, R.S., Fraser, A.G., Dong, Y., Poulin, G., Durbin, R., Gotta, M., Kanapin, A., LeBot, N., Moreno, S., Sohrmann, M., Welchman, D.P., Zipperlen, P. and Ahringer, J. (2003) Systematic functional analysis of the *Caenorhabditis elegans* genome using RNAi. *Nature* **421**:231–237.

Fodde, R. and Smits, R. (2002) A matter of dosage. *Science* **298**:761–763.

Manning, G., Whyte, D.B., Martinez, R., Hunter, T. and Sudarsanam, S. (2002) The protein kinase complement of the human genome. *Science* **298**:1912–1934.

Strausberg, R.L., Greenhut, S.F., Grouse, L.H., Schaefer, C.F. and Buetow, K.H. (2001) In silico analysis of cancer through the Cancer Genome Anatomy Project. *Trends Cell Biol.* **11**:S66–S71.

Wang, Z., Shen, D., Parsons, D.W., Bardelli, A., Sager, J., Szabo, S., Ptak, J., Silliman, N., Peters, B.A., van der Heijden, M.S., Parmigiani, G., Yan, H., Wang, T.-L., Riggins, G., Powell, S.M., Willson, J.K.V., Markowitz, S., Kinzler, K.W., Volgelstein, B. and Velculescu, V.E. (2004) Mutational analysis of the tyrosine phosphatome in colorectal cancers. *Science* **304**:1164–1166.

2

DNA structure and stability: mutations vs. repair

Introduction

Genetic information, coded within DNA, requires stability. DNA directs the production of proteins needed for the structure and function of cells over a lifetime, through an adaptor molecule, RNA. Unlike RNA and protein that have a limited existence before they are degraded and/or recycled, DNA must maintain its integrity over that lifetime. However, our genes are subject to a myriad of attacks by both environmental agents and endogenous processes that result in mutation and scission. Changes to the DNA sequence may have severe consequences for the cell and its progeny. Cancer is a disease that involves alterations to gene structure and gene expression at the cellular level. The role of the accumulation of mutations is well established for carcinogenesis. In this chapter we will review the structure of a gene and describe the mutations that occur during carcinogenesis.

When considering the process of carcinogenesis we must be aware that cells are equipped with defense mechanisms against mutations, such as the detection and repair of DNA damage. Detection and repair of DNA damage is particularly crucial in the time before a cell divides since errors existing during replication will be passed on to daughter cells. Pausing the cell cycle is sometimes coupled to the repair of DNA damage. Apoptosis, a more hard-line defense, can be triggered as a last resort; thus, cell suicide is the ultimate price paid to prevent perpetuation of DNA damage and to protect the individual from carcinogenesis (see Chapter 6). In this chapter, we will also examine how mutations in DNA occur as a consequence of exposure to carcinogens and, on the other hand, examine the DNA repair systems that are in place to maintain the integrity of the genome and suppress tumorigenesis.

2.1 Gene structure—two parts of a gene: the regulatory region and the coding region

We have 30,000 genes! They are encoded in our DNA, an impressively simplistic double-helical molecule made up of two chains of nucleotides. The nucleotide is made up of a sugar, phosphate, and a nitrogenous base (adenine, guanine, cytosine, or thymine) and it is the sequence of the bases that holds the instructional information of our genes. The central dogma of molecule biology states that DNA is transcribed into RNA and RNA is translated into protein. Gene expression refers to the transcription of a gene. For the purpose of simplicity, keep in mind that there are two distinct functional parts to a gene (Figure 2.1). The 5′-end of a gene contains nucleotide sequences that make up the promoter region and this region is involved in regulating the expression of the gene. These 5′ nucleotide sequences interact with proteins that affect the activity of RNA polymerase and determine when and where a gene is expressed. [Note however, there are exceptions; for many genes, some regulatory regions can be located elsewhere, such as downstream of the gene or within introns.] The TATA box (TATAAAA) located near the start site of transcription is one of the most important regulatory elements for most genes. Binding of the TATA box-binding protein (TBP) is crucial for the initiation of transcription. A short sequence of DNA within a promoter that is recognized by a specific protein that contributes to the regulation of the gene is called a response element (RE). Common response elements identify genes under a common type of regulation. For example, the sequence CCATATTAGG is referred to as the serum response element (SRE) and is found in genes that are responsive to serum. Also, it is not surprising that the response element for a protein that is essential for the regulation of the cell cycle, the transcription factor E2F, is found in the promoters of the cyclin E and cyclin A genes, the products of which are major players in the cell cycle. Enhancer elements are additional regulatory DNA sequences that are position and orientation-independent relative to a promoter and are important for tissue-specific and stage-specific expression. Downstream (the direction along the DNA molecule towards the 3′-end) of the promoter are the nucleotides that will be transcribed into RNA and those coding for exons will be translated into protein. These downstream nucleotide sequences represent the coding region of the gene.

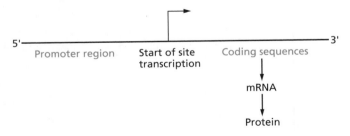

Figure 2.1 A simplistic representation: two functional parts of a gene

2.2 Mutations

As stated previously, most carcinogens are mutagens. These agents induce mutations either by modifying DNA (e.g. forming DNA adducts)
or by causing chromosomal damage (e.g. DNA strand breaks). Several
types of mutations are illustrated in Figure 2.2: transitions, transversions, insertions, deletions, and chromosomal translocations. Transitions
and transversions are two types of base substitutions. A transition is the
substitution for one purine for another purine and a transversion is the
substitution of a purine for a pyrimidine or vice versa. Base substitutions during replication may occur for several reasons. First, DNA polymerase is not always 100% accurate. The enzyme may make an error
and insert a wrong nucleotide during DNA synthesis. Also, modifications of bases due to oxidation or covalent additions and alterations of
chromatin structure can cause misreading of the DNA template by DNA
polymerase. Remember that the genetic code is a triplet code read in a sequential but non-overlapping manner. An insertion or deletion of a base
can alter the reading frame (marked by a "," in Figure 2.2) and thus can
also be referred to as a frameshift mutation. In most cases this leads to
a non-functional or truncated protein product. A chromosomal translocation is the exchange of one part of one chromosome for another part
of a different chromosome and results in changes of the base sequence
of DNA. As we will see in later chapters, there are many examples of
these types of mutations in genes regulating growth, differentiation, and
apoptosis that are involved in carcinogenesis. In theory, initial mutations may occur anywhere across a particular gene but the location will
determine whether some of these mutations give rise to a growth advantage and contribute towards carcinogenesis. For example, a mutation
may alter the conformation of a cyclin protein and result in unregulated

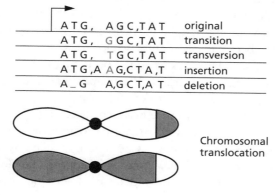

Figure 2.2 Types of mutations

progression of the cell cycle whereas another mutation may have no effect on protein conformation or function.

The consequence of a mutation in a gene is determined by its location with respect to the two functional parts of a gene. Mutations occurring in the promoter region may alter the regulation of the gene and affect the levels or temporal/spatial expression of the gene product. The consequence of such mutations may be over- or under-expression of the protein product or the appearance of the protein product at the wrong time or in the wrong place (i.e. the wrong cell type), respectively. Alternatively, mutations occurring in the coding region of genes may affect the structure and thus alter the function of the gene product or cause a truncation (e.g. introduction of a stop codon) that abolishes the protein's function completely.

2.3 Carcinogenic agents

The backbone of cancer biology has been the identification of carcinogens responsible for cancer-causing mutations, and the identification of specific mutations as causative factors of carcinogenesis along with the elucidation of the pathways they affect. Several classes of carcinogens will now be described, including radiation, chemicals, infectious pathogens, and particular endogenous reactions.

Radiation as a carcinogen

Radiation is energy. There are two forms of radiation: energy traveling in waves or as a stream of atomic particles. Energy waves include gamma (γ) rays, high-energy electromagnetic radiation that is similar to X-rays. Atomic particles include alpha (α) and beta (β) particles that are emitted by radioactive atoms. (Alpha particles comprise two protons and two neutrons, while beta particles comprise electrons.)

Electromagnetic radiation is naturally occurring radiation, which possesses a broad range of energies. Electromagnetic radiation moves as waves of energy, which have peaks and troughs (in a manner analogous to waves at sea). The distance between successive peaks (or troughs) is termed the wavelength. High-energy electromagnetic radiation such as cosmic radiation has a short wavelength, while low-energy radiation such as radio waves has a long wavelength. The electromagnetic spectrum spans electromagnetic radiation of varying wavelengths, as shown in Figure 2.3. The electromagnetic spectrum extends from long wavelength radiation (not shown) to extremely short wavelength radiation, such as X-rays and gamma radiation. The visible spectrum spans

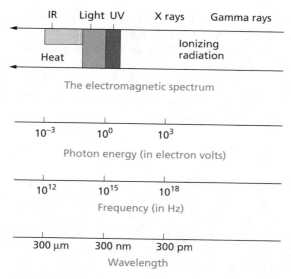

Figure 2.3 The electromagnetic spectrum and corresponding characteristics

those wavelengths that we can detect with our eyes as visible light. Ultraviolet (UV) radiation is emitted from the sun and has a higher energy (and so shorter wavelength) than visible light.

Several types of radiation (including both energy waves and atomic particles) can damage DNA and act as carcinogens. The amount of energy released by a particular radiation source affects the mechanism and extent of damage to DNA. The amount of energy released by a particular radiation source and absorbed by the body tissue is measured in grays (Gy). 1 Gy is the release to the body tissue of 1 J of energy per 1 kg of tissue. The real issue is not how much radiation is absorbed by the body tissue however, but how much damage is done when the radiation has been absorbed. The amount of damage caused depends on the rate at which a particular radiation source releases energy. If a radiation source releases energy at a high rate, then it causes more damage than a source that releases energy more slowly.

Linear energy transfer (LET) is used to help describe the rate at which energy is released. Specifically, it describes the amount of energy released by a radiation source as it travels a fixed distance. High LET radiation emits more energy than low LET radiation over the same distance. Therefore, high LET radiation (such as alpha particles) causes more biological damage than low LET radiation (such as X-rays). The quantity and type of DNA damage caused by a particular radiation source depends on whether it is high or low LET radiation. Double-stranded DNA breaks are more commonly caused by high LET radiation, and lead to chromosomal translocations and deletions.

The amount of biological damage caused by a particular source of radiation is measured in sieverts (Sv). (The numerical value of these units is determined by multiplying the gray units by a factor relating to the LET value of a particular type of radiation.)

Two classes of radiation, ionizing radiation and UV radiation, have been demonstrated to act as carcinogens and damage DNA. Let us examine both types of radiation below.

Ionizing radiation

Ionizing radiation includes both alpha and beta particles (atomic particles) and gamma rays (energy waves). When high-energy radiation, such as γ-rays, strikes molecules in its path, electrons may be displaced from atoms within the molecule. The loss of one or more electrons converts the molecule from being electrically neutral, to carrying an electrical charge. The charged molecule is called an **ion** and hence the radiation causing the formation of an ion is called ionizing radiation.

Ionizing radiation can damage DNA directly by causing ionization of the atoms comprising DNA, or indirectly by the interaction with water molecules (a process known as **radiolysis**) to generate dangerous intermediates called **reactive oxygen species (ROS)** (see Box below). These reactive oxygen species may react with DNA, or with other biomolecules, to cause damage within the cell.

A little lesson about ROS...

Some radiation exerts its biological effect by the generation of damaging intermediates through the interaction of radiation with water, or radiolysis. Since our body comprises between 55–60% water, radiation is most likely to strike water than any other matter. The striking of water by radiation causes it to loose an electron and become highly reactive. This sets off a chain reaction in which water is converted to oxygen, O_2, through a three-step process. Radiation interacts with a single molecule of water and thus it cannot split water directly into the diatomic gases H_2 and O_2. Equation (1) below is **not** possible because this equation is not balanced.

$$H_2O \rightarrow H_2 + O_2 \tag{1}$$

The balanced equation (Equation 2) does **not** apply to radiation since radiation interacts with only a single molecule of water, and not the two molecules required in Equation (2).

$$2H_2O \rightarrow 2H_2 + O_2 \tag{2}$$

Instead, radiolysis results in the sequential generation of three dangerous reactive oxygen species (ROS) as an electron (e−) is lost at each step. The three ROS, formed in sequence, are the hydroxyl radical (−OH), hydrogen peroxide (H_2O_2), and the superoxide radical ($O_2 \cdot^-$).

$$H_2O \xrightarrow{e-} \underset{\text{Hydroxyl radical}}{\cdot OH} \xrightarrow{e-} \underset{\text{Hydrogen peroxide}}{H_2O_2} \xrightarrow{e-} \underset{\text{Superoxide radical}}{O_2 \cdot^-} \xrightarrow{e-} O_2$$

The hydroxyl radical is an extremely reactive molecule; in fact, it is one of the most reactive (and therefore dangerous) molecules known! It immediately removes electrons from any molecule in its path, turning that molecule into a free radical and so propagating a chain reaction. (A free radical is a highly unstable, reactive molecule that possesses an unpaired electron. Both the hydroxyl radical and superoxide radical shown above are free radicals.)

Neither hydrogen peroxide nor the superoxide radical are as reactive as the hydroxyl radical. Hydrogen peroxide is actually more dangerous to DNA than the hydroxyl radical however. The slower reactivity of hydrogen peroxide (compared with the hydroxyl radical) gives the hydrogen peroxide molecule time to travel into the nucleus of a cell, where it is free to interact with and wreak havoc upon DNA.

Oxidation of DNA (the removal of electrons, by species such as the free radicals mentioned here) is one of the main causes of mutation, and explains why free radicals are such potent carcinogens. Oxidation can produce several types of DNA damage, including oxidized bases. Among the variety of oxidized nitrogenous bases observed, 8-oxo-guanine is the most abundant. DNA polymerase mispairs 8-oxo-guanine with adenine during DNA replication leading to a G → T transversion mutation. The presence of iron can exacerbate the consequences of H_2O_2 production. If it encounters iron and receives an electron from it, hydrogen peroxide can be reconverted into the hydroxyl radical that may attack DNA. The Fenton reaction (Equation 3) illustrates this:

$$H_2O_2 + Fe^{2+} \rightarrow OH^- + \cdot OH + Fe^{3+} \qquad (3)$$
$$O_2 \cdot^- + Fe^{3+} \rightarrow O_2 + Fe^{2+} \qquad (4)$$

The superoxide radical is the third intermediate before the formation of oxygen. It is not very reactive but acts more as a catalyst for the generation of the other two intermediates mentioned because it helps regenerate iron (Equation 4) in the form needed for the above-mentioned Fenton reaction. Thus, the ROS intermediates affect one another.

(Lane, 2002.)

People are exposed to varying amounts of ionizing radiation. Exposure to gamma rays from cosmic radiation depends on the altitude at which you live or travel. The average exposure for high altitude flights is about 0.005–0.01 mSv per hour. A chest X-ray required for medical diagnosis of some conditions exposes patients to 0.1 mSv. The contribution of the accumulation of these varying daily exposures towards cancer risk is relatively unknown.

Studies of the victims of the atomic bombing in Japan continue to contribute to our knowledge of ionizing radiation as a carcinogen. Evidence suggests that the most important damage associated with ionizing

PAUSE AND THINK

Marie Curie, who worked with radioactivity all of her adult life, died of leukemia at the age of 67. The multi-stage process of carcinogenesis is manifested by the fairly long incubation periods needed to develop cancer after exposure to carcinogens.

LIFESTYLE TIP

Short-term tanning salon exposure (10 treatments in 2 weeks) results in cyclobutane pyrimidine dimer formation. Look at the data in Whitmore *et al.* (2001). A natural look is less taxing for your DNA!

radiation-induced carcinogenesis is double-strand DNA breaks. The Radiation Effects Research Foundation publishes periodic reports on the mortality of The Life Span Study cohort of 80,000 atomic bomb survivors (latest report: Preston *et al.*, 2003). These studies have revealed three important points: 1) leukemia is the most frequent ionizing radiation-induced cancer, 2) age is an important risk factor, whereby those exposed as children are most affected and 3) risks of solid cancer increase with dose in a linear fashion. People exposed at 30 years of age have a risk of solid cancer that is elevated by 47% per sievert at age 70.

Ultraviolet radiation

Ultraviolet radiation (UV) from the sun is also carcinogenic and is a principal cause of skin cancer. Of the three types of UV light: UVA (wavelength 320–380 nm), UVB (wavelength 290–320 nm), and UVC (wavelength 200–290 nm), UVB is the most effective carcinogen. The conjugated double bonds in the rings of the nitrogenous bases of DNA absorb UV radiation. UVB directly and uniquely causes characteristic UV photoproducts: cyclobutane pyrimidine dimers and pyrimidine-pyrimidone photoproducts (Figure 2.4a,b). Cyclobutane pyrimidine dimers are most prevalent, formed at least 20–40 times more frequently than other UV photoproducts. The formation of a pyrimidone (6–4) photoproduct mimics an abasic site (a nucleotide minus a base) and is more efficiently repaired than cyclobutane pyrimidine dimers. The formation of a pyrimidine dimer causes a bend in the DNA helix and, as a result, DNA polymerase cannot read the DNA template. Under these conditions DNA polymerase preferentially incorporates an "A" residue. Consequently, TT dimers are often restored but TC and CC dimers result in transitions (TC → TT and CC → TT) (Figure 2.4c). Results from a mammalian cell system showed that cyclobutane pyrimidine dimers are responsible for at least 80% of UVB-induced mutations. The precise class of mutations resulting from pyrimidine dimers is a unique molecular signature of skin cancer (see Box on Skin Cancer below); they are not found in any other types of cancer.

UVA indirectly damages DNA via free radical-mediated damage. Water is fragmented by UVA, generating electron-seeking ROS (such the hydroxyl radical as mentioned above) that cause DNA damage (e.g. oxidation of bases). G → T transversions are characteristic of UVA damage.

Skin cancer

UV light is specifically carcinogenic to the skin because it does not penetrate the body any deeper than skin. The skin is made up of squamous cells, basal cells, and

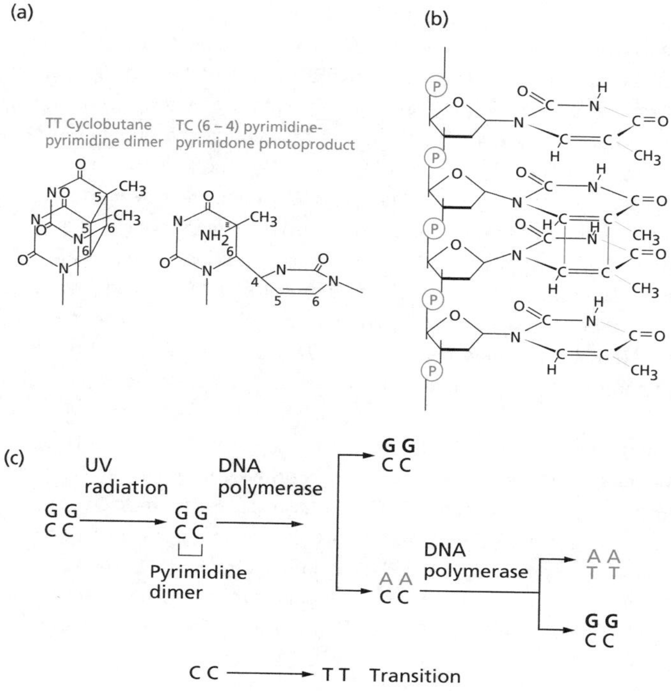

(a)

(b)

TT Cyclobutane
pyrimidine dimer

TC (6 – 4) pyrimidine-
pyrimidone photoproduct

(c)

C C ⟶ T T   Transition

Figure 2.4 (a) UV photoproducts; (b) A pyrimidine dimer in the context of a polynucleotide chain; (c) Steps involved in UV induced transitions

melanocytes and skin cancers are classified by the cell type they affect: squamous cell carcinoma (SCC), basal cell carcinoma (BCC) and melanoma respectively. The depth of transmission of each type of UV light is dependent on the wavelength: UVC only penetrates into the superficial layer of the skin, UVB penetrates into the basal level of the epidermis and UVA penetrates into the more acellular dermis level. Sunscreens work on the basis of including UV-absorbing organic chemicals (e.g. cinnamates), inorganic zinc-containing pigments, or titanium oxides in their ingredients to minimize UV absorption by the skin. (Note that melanin formation, known to most people as tanning, is a natural defense mechanism against UV absorption.) Additional ingredients in sunscreens must be used with care as we have learned that some compounds may be photosensitized carcinogens, chemicals that can be activated by UV to become carcinogenic. Ironically, some early sunscreens included bergamot oil which contains 5-methoxy psoralen, a photosensitized carcinogen! Some drugs such as fluoroquinolone antibiotics are also photosensitized carcinogens and explains the reasons for the precautions from doctors to stay out of the sun during their administration.

A cellular mechanism for the elimination of UV damaged skin cells is to initiate apoptosis. This phenomenon is familiar to us as the peeling of the skin after a sunburn. The tumor suppressor p53 protein (introduced in Chapter 1 and discussed in detail in Chapter 5) is an important regulator of apoptosis. Mutation of the *p53* gene is important for the initiation of squamous cell and basal cell carcinoma, but not melanoma. The characteristic mutations (CC → TT transitions) caused only by UV and no other carcinogen were identified in the *p53* gene. Mutations in the *p53* gene which disrupt normal p53 function and provide cells with a growth advantage, may induce the formation of tumor cells. The pattern of mutation is not random but rather tends to be localized to nine places, called hot spots. This suggests that *p53* mutations are causal for skin cancer. Further investigation of why there are so few hotspots within the context of hundreds of sites with adjacent pyrimidine dimers in the *p53* gene yielded an explanation. The hot spots in *p53* are not repaired efficiently. Removal of cyclobutane pyrimidine dimers is particularly slow at these sites. The resulting loss of p53 function causes a block in apoptosis and consequently allows the proliferation of mutated *p53* cells. Thus UV radiation not only induces *p53* mutations but also selects for the clonal expansion of the *p53* mutated cells, by inducing apoptosis in normal cells with wild-type *p53*.

Different pathways seem to be central for melanoma. The elucidation of one of these pathways was one of the first successes of the Cancer Genome Project. It identified mutations in the *BRAF* gene in 66% of malignant melanomas (Davies *et al.*, 2002). BRAF is a serine/threonine kinase that functions in the signal transduction pathway downstream of α melanocyte stimulating hormone and may explain why there is a high frequency of *BRAF* mutations in melanoma relative to other cancers. Surprisingly, the major mutation identified (T → A) in the kinase domain is not characteristic of UV-induced mutations (CC → TT).

Chemical carcinogens

Many chemicals in our environment and in our diet play a role in human carcinogenesis. The common mechanism of action of carcinogens is that an electrophilic (electron deficient) form reacts with nucleophilic sites (sites that can donate electrons) in the purine and pyrimidine rings of nucleic acids. Some chemical carcinogens can act directly on DNA but others become active only after they are metabolized in the body, forming what are called ultimate carcinogens, the molecules that execute the damage. A family of enzymes, called the cytochrome P450 enzymes, is involved in the metabolism of chemicals in the liver and is important in the activation of carcinogens to ultimate carcinogens. Genetic polymorphisms and variable expression account for differences in responses to chemical carcinogens among individuals. For example, the expression of one of the P450 enzymes called CYP1A1 (aryl hydrocarbon hydroxylase) can vary 50-fold in human lung tissue and may be responsible for the delivery of varying doses of ultimate carcinogens among smokers (Alexandrov *et al.*, 2002).

Carcinogens can be segregated into ten groups:

(i)   polycyclic aromatic hydrocarbons

(ii)   aromatic amines

(iii) azo dyes

(iv)   nitosamines and nitosamides

(v)   hyrazo and azoxy compounds

(vi)   carbamates

(vii)   halogenated compounds

(viii)   natural products

(ix)   inorganic carcinogens

(x)   miscellaneous compounds (alkylating agents, aldehydes, phenolics)

Four major classes of carcinogens are described below: polycyclic aromatic hydrocarbons (PAHs), aromatic amines, nitrosamines, and alkylating agents. These carcinogens exert their effects by adding functional groups covalently to DNA. Chemically modified bases, called DNA adducts, distort the DNA helix causing errors to be made during replication. The resulting mutations initiate cell carcinogenesis.

Polycyclic aromatic hydrocarbons (PAH)

The first demonstration that chemicals could be used to induce cancer in animals was carried out in 1915. Coal tar, containing carcinogenic polycyclic aromatic hydrocarbons (PAHs), induced skin carcinomas on the ears of rabbits. Carcinogenic PAHs are derived from phenanthrene (Figure 2.5a). Additional rings and/or methyl groups in the bay region of the three aromatic rings can convert inactive phenanthrene into an active carcinogen. Benzo[a]pyrene (BP), the most well-known carcinogen in cigarette smoke, and 7,12-dimethyl benzanthracene (DMBA), one of the most potent carcinogens, are examples of PAHs. PAHs must be metabolized further in order to form the ultimate carcinogen that will form adducts with purine bases of DNA. The P-450 enzyme, CYP1A1, is the predominant enzyme that metabolizes BP to the highly reactive mutagenic BP diol epoxides (Figure 2.5b). BP results mainly in G to T transversions.

> **LIFESTYLE TIP**
>
> The International Agency for Research on Cancer (IARC) has classified 81 compounds in cigarette mainstream smoke as carcinogens (Smith *et al.*, 2003). BP ranks high in the measure of lipophilicity (a feature that allows easy entry into cells) and associated carcinogenicity. The presence of nicotine makes smoking addictive. Smoking is a cause of cancer; not smoking prevents illness. Don't choose to smoke.

A leader in the field of molecular carcinogenesis: Gerd Pfeifer

Gerd Pfeifer has made important contributions to determining the molecular mechanisms of cancer. Investigations into skin and lung cancer provided strong evidence that UV radiation and carcinogens in cigarette smoke are causative agents for each cancer, respectively. Gerd and his colleagues demonstrated that mutational hotspots of the *p53* gene observed in skin cancer cells is due to low efficient repair of DNA at these sites as discussed in the Box: Skin Cancer above. By mapping DNA adducts of the *p53* gene that are formed after exposure to benzo[a] pyrene diol epoxide (a potent cigarette carcinogen), Gerd and colleagues showed that the locations of these adducts matched →

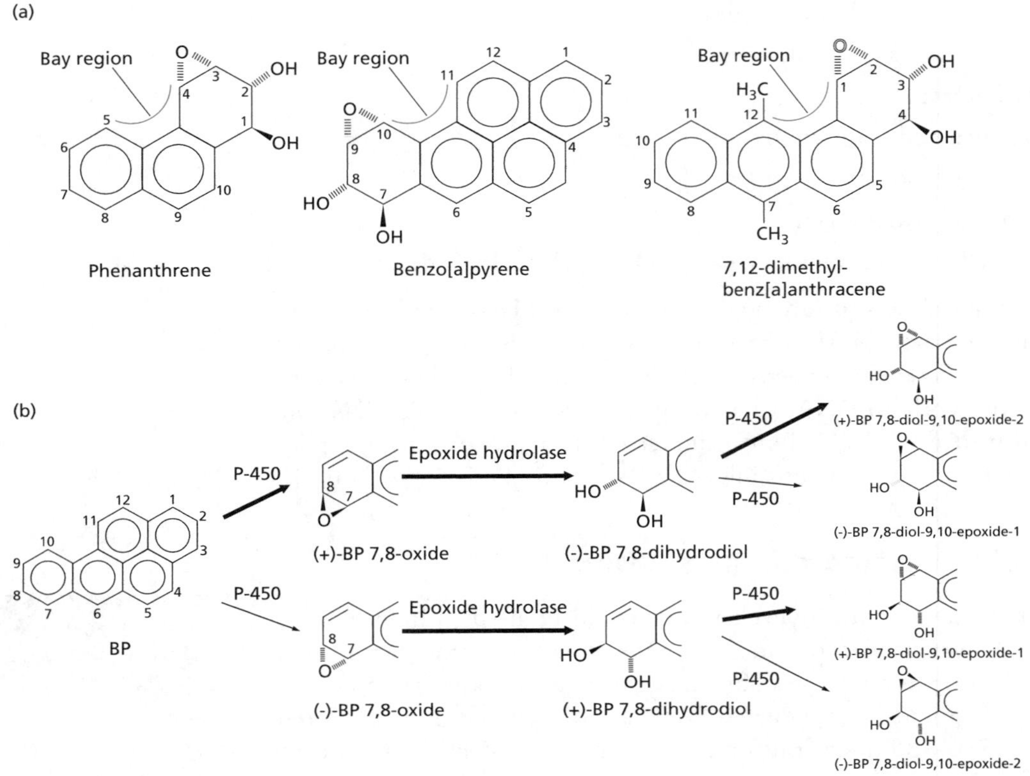

(a)

Phenanthrene

Benzo[a]pyrene

7,12-dimethyl-
benz[a]anthracene

(b)

Figure 2.5 (a) Examples of
polycyclic aromatic amines;
(b) Metabolic activation of BP

→ the distribution of *p*53 gene mutations in lung tumors from smokers. This seminal work, reported in *Science* in 1996, provided a direct causal link between a defined carcinogen and lung cancer.

Gerd Pfeifer received his Ph.D. from the University of Frankfurt, Germany. He has crossed the Atlantic and is currently a Professor and Chair at the City of Hope, Beckman Research Institute in California. His research group is continuing to study the mechanisms of mutagenesis in cancer and is also currently investigating epigenetic mechanisms of gene regulation in cancer (discussed in Chapter 3).

Aromatic amines

Heterocyclic amines (HCAs) are carcinogens produced by cooking meat, formed from heating amino acids and proteins. About 20 HCAs have been identified. Three examples, Phe-P-1, IQ, and Mel Q, are shown in Figure 2.6.

It is important to be aware of these since they illustrate an example of carcinogens, which we may be exposed to daily and are produced in our own kitchens.

PAUSE AND THINK

What is the structural difference between IQ and Mel Q?

LIFESTYLE TIP

It has been suggested that changes in the way we prepare food can reduce the amounts of HCAs produced. Oven-roasting, marinating, and coating food with breadcrumbs before frying are modifications that may reduce the formation of HCAs.

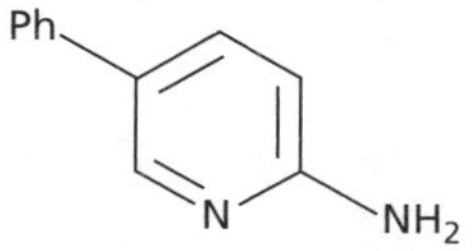

2 Amino-5 phenylpyridine (PHe-P-1)

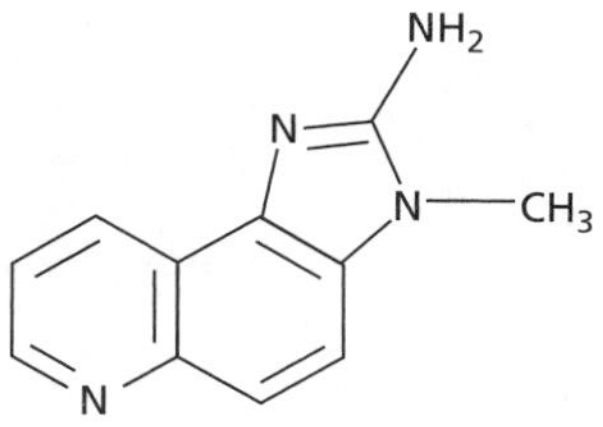

2-Amino-3 methylimidazo [4,5-f] quinoline (IQ)

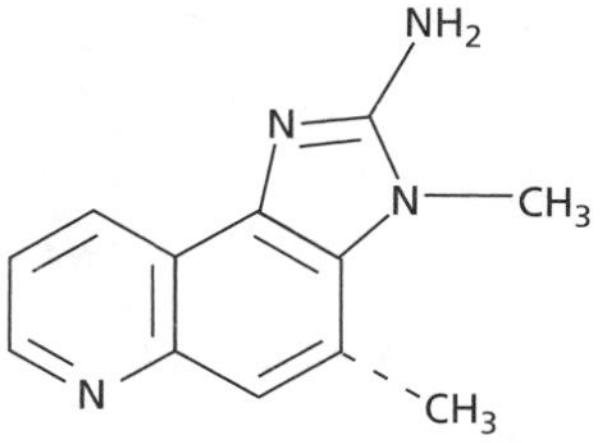

2-Amino-3,4 dimethylimidazo [4,5-f] quinoline (Mel Q)

Figure 2.6 Heterocyclic amines

Nitrosamines and nitrosamides

Many nitrosamines and nitrosamides are found in tobacco or are formed when preservative nitrites react with amines in fish and meats during smoking. The structure of alkylnitrosoureas, examples of nitrosamines, is shown in Figure 2.7a. Their principal carcinogenic product is alkylated O^6 guanine derivatives, as shown in Figure 2.7b (guanine is depicted next to it for comparison).

Alkylating agents

Mustard gas (sulfur mustard, Figure 2.8) is the most well-known example of an alkylating agent because of its use and consequences observed during World War I. It is a bifunctional (having two reactive groups) carcinogen that is able to form intrachain and interchain cross-links on DNA directly.

Infectious pathogens as carcinogens

Early in the 20th century, viruses were shown to cause tumors in animals. As we will see in Chapter 4, they have been invaluable tools for investigating the molecular events of cell transformation. Viruses that are oncogenic can be classified as DNA tumor viruses or RNA

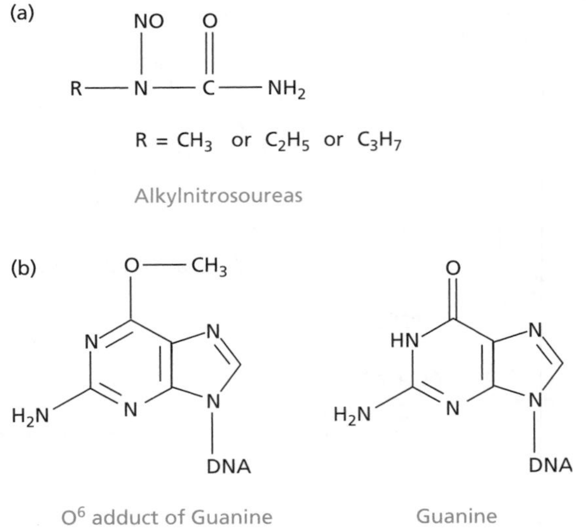

(a)

R = CH$_3$ or C$_2$H$_5$ or C$_3$H$_7$

Alkylnitrosoureas

(b)

O^6 adduct of Guanine   Guanine

Figure 2.7 (a) An example of nitrosamines: alkylnitrosoureas; (b) A potential carcinogenic product of nitrosamines: O^6 adduct of guanine. Guanine is shown for comparison

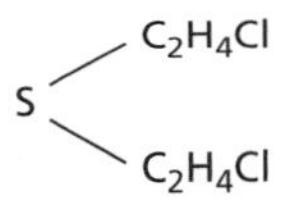

Figure 2.8 Structure of mustard gas

tumor viruses (also called retroviruses), depending on the nucleic acid that defines their genome. The mechanisms of carcinogenesis for these two classes of viruses differ. DNA tumor viruses encode viral proteins that block tumor suppressor genes, often by protein–protein interactions (discussed in Chapter 5). Retroviruses encode mutated forms of normal genes (i.e. oncogenes) that have a dominant effect in host cells (discussed in Chapter 4). Their mechanisms of replication also differ. Some DNA viruses, such as human papilloma and Epstein-Barr viruses, replicate strictly as episomes within host cells. Retroviruses replicate by integration of the viral genome into the host DNA and utilize the host's translational machinery to produce viral proteins. Integration may lead to deregulated gene expression.

Studies have shown direct causal effects of viruses in human cancer. The International Agency for Research on Cancer (IARC) has classified human papillomavirus (type 16 and 18) as a human carcinogen and a causative agent of cervical cancer. Human papillomavirus DNA has been detected in virtually all cervical cancers worldwide. Cervical cancer accounts for 5% of all cancers worldwide and thus eradication of human papillomavirus infection would prevent most of these cases (see Chapter 10). In addition, hepatitis B virus is associated with liver cancer, and Epstein-Barr virus (EBV) with lymphomas. The human T-cell leukemia virus type 1 (HTLV-1) is the only retrovirus known directly to cause cancer in humans.

More recently, the involvement of bacteria in human cancer has been recognized. *Helicobacter pylori*, a gram-negative spiral bacterium, establishes chronic infection and ulcers in the stomach and alters host cell function which is associated with carcinogenesis. The International Agency for Research on Cancer has classified *H. pylori* as a human carcinogen and one of the causative agents of gastric cancer. Strong epidemiological evidence supports its carcinogenic role, although *H. pylori* accounts for fewer than 50% of gastric cancer cases in Western societies (Kelley and Duggan, 2003). One study noted by Kelley and Duggan (2003) has reported that the risk of gastric cancer after *H. pylori* infection increased nine-fold after a 15 year or more follow-up. The typhoid pathogen, *Salmonella enterica serovar Typhi* (*S. typhi*), establishes chronic infection in the gall bladder and has been linked to hepatobiliary and gall-bladder carcinoma. The molecular events behind the mechanism of bacteria-induced transformation are the subject of present studies. The promotion of host cell proliferation, the generation of oxygen free radicals and subsequent DNA damage, and the activation of oncogenes are areas of investigation (Lax and Thomas, 2002).

Endogenous carcinogenic reactions

In addition to carcinogens, endogenous cellular reactions generate mutations. Oxidative respiration and lipid peroxidation, two processes of normal cell metabolism, produce ROS that can react with DNA and lipids to produce oxidized products (e.g. 8-oxo-guanine) also seen by exposure to radiation (see above). During respiration, the initiating radical, superoxide anion ($O_2 \cdot^-$) is produced upon reduction of NADH and formation of ubisemiquinone during oxidative phosphorylation. Therefore, breathing generates the same ROS intermediates as those generated by radiation! However, the dose of these intermediates differs between the two sources: radiation produces extremely reactive hydroxy radicals immediately and randomly within a cell, while respiration produces the less reactive superoxide radical immediately and only at specific locations within the cell.

Spontaneous chemical reactions (e.g. hydrolysis of the glycosidic bond between a base and deoxyribose producing an abasic site) also contribute to the formation of mutations. Deamination of cytosine to form uracil is the most common. Errors during DNA replication and DNA recombination contribute to the formation of mutations although the DNA polymerases used possess proofreading ability to help minimize the amount of mutations caused in this way. The proofreading function is dependent on the 3'-5' exonuclease activity of the polymerase. If an incorrect nucleotide is added to the growing 3'-end of the newly synthesized strand the DNA double helix exhibits melting; that is, the strands remain separated at this point. Melting causes the polymerase to pause and the strand is transferred to the exonuclease site. Here the incorrect nucleotide is removed, the strand is transferred back to the original polymerase binding site, and DNA synthesis reoccurs. Overall, it is estimated that 10^4 to 10^6 mutations occur in a single human cell per day. By and large, this immense error burden is successfully dealt with by the highly efficient cellular DNA repair mechanisms under normal circumstances.

2.4 DNA repair and predispositions to cancer

DNA repair is an important line of defense against mutations caused by carcinogens and by endogenous mechanisms. If DNA lesions are not repaired before a cell replicates, they may contribute to carcinogenesis. Repair of the variety of mutations is accomplished by several different DNA repair mechanisms. Five types of DNA repair systems are described below: one-step repair, nucleotide excision repair, base excision repair, mismatch repair, and recombinational repair. Defects in most of these pathways result in the predisposition to cancer.

One-step repair

One-step repair involves the direct reversal of DNA damage. The repair enzyme alkyltransferase directly removes an alkyl group from the O^6 atom of guanine after exposure of DNA to alkylating carcinogens such as N-methylnitrosourea. In this case, a methyl group is transferred to a cysteine residue on the alkyltransferase and the alkyltransferase becomes inactive.

Nucleotide excision repair (NER)

Nucleotide excision repair (NER) is specific for helix-distorting lesions such as pyrimidine dimers and bulky DNA adducts induced by environmental agents (UVB and PAHs respectively). This damage interferes with transcription and replication as described above. Two subpathways exist: global genome NER surveys the genome for helix-distortion and transcription-coupled repair identifies damage that interferes with transcription. The lesion, along with some (24–32) adjacent nucleotides, are excised out by endonucleases, and DNA polymerase δ/ε is used to fill in the gap using the opposite strand as a template. Proliferating cell nuclear factor is part of the polymerase holoenzyme and physically forms a ring that encircles and binds the damaged region. Xeroderma pigmentosum

(XP) is an inherited disorder characterized by a defect in NER. Affected individuals are hypersensitive to the sun and have a 1000-fold increased risk of skin cancer. Seven XP gene products (XPA–XPG) have been identified out of the 25 proteins involved in NER.

Base excision repair

Base excision repair targets chemically altered bases induced mostly by endogenous mechanisms; in the absence of such repair the damage will cause a point mutation. The chemically altered bases may be small enough not to interfere with replication or transcription. Glycosylases flip the lesion outside of the helix and cleave the base from the DNA backbone creating an abasic site. Subsequently an endonuclease cleaves the DNA strand at the abasic site and DNA polymerase β replaces the nucleotide and ligase fills the gap. Mutations in the *OGG1* gene that codes for the principal glycosylase responsible for the repair of 8-oxoguanine have not been identified in tumors to date. No inherited defects in BER had been identified in humans until recently—mutations in the *MYH* gene that encodes a DNA glycosylase responsible for the removal of mismatched adenines paired with 8-oxoguanine may be the principal cause of multiple colorectal adenoma syndrome (Al-Tassan *et al.*, 2002).

Mismatch repair

Mismatch repair corrects replication errors that have escaped editing by polymerases. It includes repair of insertions and deletions produced as a result of slippage during the replication of repetitive sequences as well as nucleotide mismatches. The molecular events are described in brief:

- recognition of the mismatch is carried out by proteins HMSH2/6 and hMSH2/3
- hMLH1/hPMS2 and hMHL1/hPMS1 are recruited
- the newly synthesized strand is identified (flagged by the replication machinery)
- endonucleases and exonucleases remove the nucleotides around and including the mismatch
- DNA polymerases resynthesize a newly replicated strand

Hereditary non-polyposis colorectal cancer (HNPCC) is one of the most common cancer syndromes in humans. Half of all patients with hereditary non-polyposis colorectal cancer carry a germline mutation in *hMLH1* or *hMSH2*. Loss of function of the protein products encoded by these genes is responsible for complete loss of mismatch repair. Thus, cells are vulnerable to mutations.

Recombinational repair

Homologous recombination and non-homologous end-joining are two types of recombinational repair that mend double-strand DNA breaks. Homologous recombination depends on the presence of sister chromatids formed during DNA synthesis as a template for recombining severed ends. Many members of the same proteins make up a complex that performs what has been nicknamed DNA gymnastics. The molecular events shown in Figure 2.9 are described in brief:

(a) A double-strand break activates the ataxia telangiectasia mutated (ATM) kinase.

(b) The RAD50/MRE11/NBS1 complex (a substrate of ATM) uses its 5′-3′ exonuclease activity (depicted by scissors in Figure 2.9) to create single-stranded 3′-ends.

(c) BRCA2 aids in the nuclear transport of RAD51 (shown as round gray circles).

(d) RAD52 facilitates RAD51 binding to these exposed ends to form a nucleoprotein filament.

(e) RAD51 can exchange a homologous sequence from a single strand within a double-stranded molecule (shown in red; e.g. a sister chromatid), with a single stranded sequence.

(f) The sequences from the double-stranded molecule are then used as a template sequence for repair.

(g) Resolvases restore the junctions formed as a result of homologous recombination, called Holliday junctions.

(h) Two copies of intact DNA molecules are produced with rarely any errors.

Ataxia telangiectasia is an inherited syndrome whereby patients have a mutation in the ataxia telangiectasia mutated (ATM) kinase. Patients are sensitive to X-rays and have an increased risk of lymphoma.

Suffice it to say that the other type of recombinational repair, end-joining, links non-homologous ends and is therefore error prone, and can possibly result in chromosomal translocations.

One of the main molecular players involved in carcinogenesis is p53 and should be mentioned here. p53, "the Guardian of the Genome", is a protein that plays an important role in the molecular events that protect the integrity of DNA; it is central in the orchestration of DNA repair. The details of this important tumor suppressor protein will be discussed in Chapter 5.

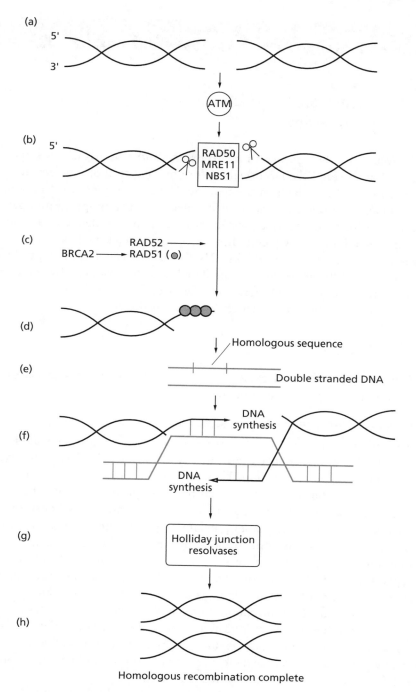

Figure 2.9 Recombinational repair

2.5 Conventional therapies: chemotherapy and radiation therapy

Conventional therapies continue to extend and save lives. It is important to understand their rationale before moving to more molecular approaches discussed later in the text. Several conventional therapies aim to induce extensive DNA damage in order to trigger apoptosis and paradoxically include agents classified as carcinogens. Other conventional therapies inhibit DNA metabolism in order to block DNA synthesis in the rapidly dividing cancer cells. DNA synthesis is essential to produce a new set of chromosomes for the daughter cells produced by cell division. Still other drugs interfere with the mechanics of cell division. Both chemotherapies and radiotherapy are described below.

Chemotherapy

A brief description and examples of the three main types of classical chemotherapy are discussed below.

Alkylating agents and platinum-based drugs

Alkylating agents and platinum-based drugs work by a similar mode of action. Alkylating agents have the ability to form DNA adducts by covalent bonds via an alkyl group. They may act during all phases of the cell cycle. Chlorambucil (Figure 2.10a) is one example of a member of the nitrogen mustard family of drugs. Its usual targets are the N7 position of guanine residues. Bifunctional alkylating agents (compounds with two reactive groups) form intra-strand and inter-strand cross-links in DNA that alter the conformation of the double helix or prevent DNA strand separation and interfere with DNA replication. They are much more potent than monofunctional analogs indicating that cross-linking is the basis of their function since monofunctional analogs cannot cross-link.

Some drugs require metabolic activation within the body. The alkylating agent cyclophosphamide (Figure 2.10b) is one example. Oxidases in the liver produce an aldehyde form that decomposes to yield phosphoramide mustard, the biologically active molecule.

The platinum-based drugs, such as cisplatin [cis $Pt(II)(NH_3)_2Cl_2$] and carboplatin (Figure 2.10c and 2.10d, respectively), form covalent bonds via the platinum atom. Cisplatin is a water-soluble molecule that contains a Pt atom bound to four functional groups. The Pt-N bond has

PAUSE AND THINK

Do you recall a similar mechanism of action for any carcinogens?

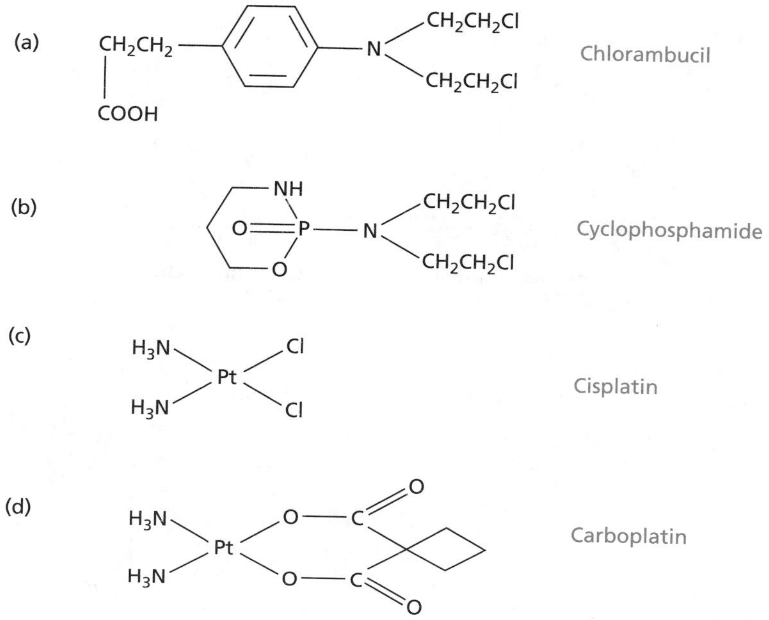

Figure 2.10 Examples of alkylating agents and platinum-based drugs

covalent character and is essentially irreversible whereas that with Cl is more labile. Cl is replaced with water in the plasma and cytosol before the molecule binds to the N7 position of guanine and adenine in its DNA target. The GG, AG, and GXG (where X can be any base) adducts comprise over 90% of the total. The resulting DNA damage triggers apoptosis. Although cisplatin had a major impact on some cancers, such as ovarian cancer, it was associated with irreversible kidney damage. Later, carboplatin was identified as a less toxic platinum analog.

Antimetabolites

Antimetabolites are compounds that are structurally similar to endogenous molecules (e.g. nitrogenous bases of DNA) and therefore can mimic their role and inhibit nucleic acid synthesis. Two examples, fluorodeoxyuridylate and methotrexate, are shown alongside similar endogenous molecules, deoxyuridylate and tetrahydrofolate respectively, in Figure 2.11. 5-Fluorouracil (5-FU) is a derivative of uracil and is converted into fluorodeoxyuridylate (F-dUMP) *in vivo*. F-dUMP competes with the natural substrate dUMP for the catalytic site of thymidylate synthase, the enzyme that produces thymidylate (dTMP) (Figure 2.12). F-dUMP forms a covalent complex with the enzyme and acts as a suicide inhibitor, generating an intermediate that inactivates the thymidylate synthase through covalent modification. As a result, the dTMP and

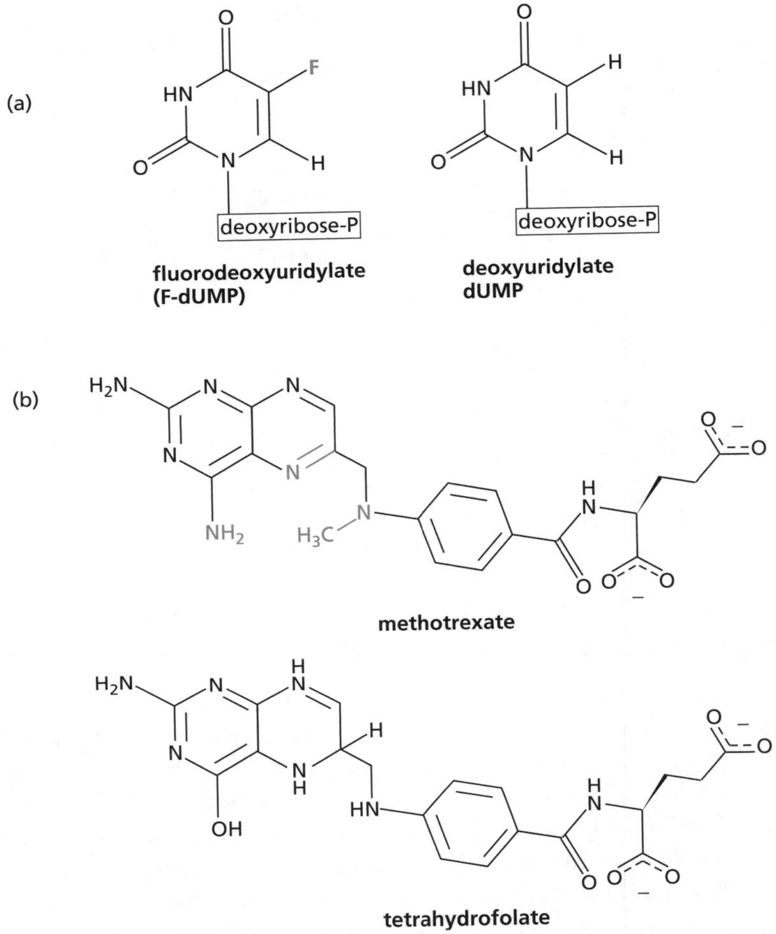

Figure 2.11 Antimetabolites: fluorodeoxyuridylate and methotrexate. Structural differences between the antimetabolite and endogenous molecule are shown in red

dTTP pools are depleted, dUMP and dUTP accumulate, and DNA synthesis in rapidly dividing cells is severely compromised. Another important antimetabolite, methotrexate, targets an accessory enzyme of the same reaction. As an analog of dihydrofolate, methotrexate is a competitive inhibitor of dihydrofolate reductase, the enzyme used to regenerate tetrahydrofolate that is required in the thymidylate synthase reaction (Figure 2.12; see Chapter 9 for further discussion of tetrahydrofolate).

Organic drugs

Doxorubicin is a fungal anthracycline antibiotic that inhibits topoisomerase II, an enzyme that releases torsional stress during DNA replication, by trapping single-strand and double-strand DNA intermediates. Doxorubicin diffuses across cell membranes and accumulates in most cell

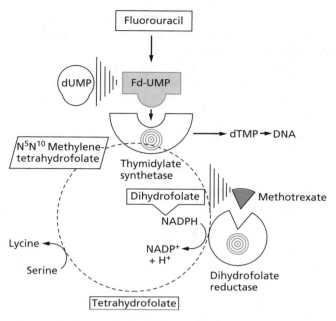

Figure 2.12 Action of antimetabolites F-dUMP and methotrexate. The enzyme thymidylate synthetase uses N^5N^{10} methylenetetrahydrofolate as a methyl donor and catalyzes the methylation of dUMP to form dTMP. The cancer drug fluorouracil is converted into the antimetabolite F-dUMP (red rectangular shape), which competes (/////) with dUMP and targets thymidylate synthetase (target symbol shown). Methotrexate (red triangle) is an antimetabolite that competes (////) with dihydrofolate and methotrexate targets the enzyme dihydrofolate reductase (target symbol shown)

types. Cardiac damage is its most severe side effect, but new compounds (e.g. ICRF-187) that can block the cardiac toxicity are being investigated. These drugs are primarily used to treat solid tumors (e.g. of the breast or lung).

The plant alkaloids vincristine and vinblastine (from the Madagascar periwinkle plant) bind to tubulin and prevent microtubule assembly in contrast to the drug paclitaxel which binds to the β-tubulin subunit in polymers and stabilize the microtubules against depolymerization. Thus two opposing strategies can be used to disrupt the mitotic spindle.

Radiation therapy

Radiation therapy, either alone or in combination with other therapies, is received by approximately 60% of cancer patients in the USA. Ionizing radiation is usually delivered to the tumor by electron linear accelerators. Radiation reacts with water inside cells to generate reactive oxygen species that damage DNA. Apoptosis will be induced in cells that contain large amounts of DNA damage. The supply of oxygen affects the potency of ionizing radiation and is thought to be due to the generation of ROS. Oxygen can assist in making radiation-induced damage permanent. More double-strand breaks occur in cells irradiated in the presence of oxygen than in cells irradiated in the absence of oxygen. Therefore, the quantity of zones of hypoxia within a solid tumor influences the outcome of radiation treatment. Targeting of the tumor has been made more precise by modern techniques such as magnetic resonance imaging (MRI) and computed tomography (CT) which produce three-dimensional images of the tumor within the body.

Heterogeneous cell sensitivity and drug resistance: obstacles to these treatments

In addition to the severe side effects that result, there are also practical problems with conventional therapies. Cancer cells, as part of a large tumor mass, will receive different doses of treatment depending on the location of individual cells within the mass. Cells deep within the tumor and therefore furthest from the blood supply will receive lower doses than cells on the surface of the tumor. Cells within the same tumor may have acquired different mutations and some cells may have become resistant to the drug.

Anticancer drugs impose a strong force for the selection of cells that can acquire drug resistance. There are several mechanisms that a cancer cell may utilize to become resistant to chemotherapy (Figure 2.13). Cells may become resistant by increasing the efflux of the drug, decreasing the intake of the drug, increasing the number of target molecules within the cell, or altering drug metabolism or DNA repair processes. Increasing the efflux of a drug is regulated at the cell surface. There is a family of ATP-dependent transporters that are involved in the movement of nutrients and other molecules across membranes. The multi-drug resistance gene (*MDR1*) codes for one member of this family called P-glycoprotein (P-gp) or the multi-drug transporter. This protein, normally a chloride ion efflux pump, can bind a variety of chemotherapeutic drugs including doxorubicin, vinblastine, and taxol. Upon binding, ATP is hydrolyzed and causes a conformational change of P-gp. As a result, the drug is released extracellularly. The transporter can be recycled by a second hydrolysis of ATP and continue to increase the efflux of the drug. Some drugs utilize specific transporters to enter cells. Mutations in these receptors may render them non-functional and decrease influx of the drug. Resistance to methotrexate commonly occurs by mutation of the folate transporter. An increase in the number of drug target molecules by gene amplification is another means of developing resistance against methotrexate. The *DHFR* gene is amplified in some cancer cells. An increase in the efficiency of DNA repair, such as increased alkyltransferase activity,

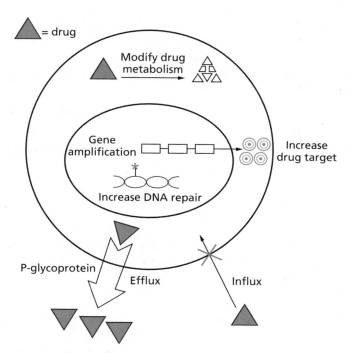

Figure 2.13 Mechanisms of drug resistance. Based on diagram printed in *Ann. Rev. Med. Vol. 53*, Copyright (2002) by Annual Reviews **www.annualreviews.org**

can give rise to resistance from alkylating agents such as doxorubicin. Levels of this enzyme are found to be highly variable in different tumors.

CHAPTER HIGHLIGHTS—REFRESH YOUR MEMORY

- Most carcinogens are mutagens.

- Several types of mutations include: base substitutions (transitions and transversions), frame-shift mutations (insertions or deletions), and chromosomal translocations.

- Mutations in the promoter region of a gene may alter its regulation.

- Mutations in the coding region of a gene may alter the function of the gene product.

- Carcinogens include radiation, chemicals, and infectious pathogens.

- Radiation can damage DNA directly or indirectly through the formation of reactive oxygen species.

- Three intermediate ROS formed from the radiolysis of water are the hydroxyl radical, hydrogen peroxide, and the superoxide radical.

- The hydroxyl radical is one of the most reactive substances.

- Many carcinogens need to be metabolized to form an ultimate carcinogen that covalently binds to DNA.

- Many chemical carcinogens add functional groups covalently to DNA.

- Both viruses and bacteria have been classified as carcinogens for specific cancers.

- One step repair, nucleotide excision repair, base-excision repair, mismatch repair, and recombinational repair are five systems for repairing damaged DNA.

- Patients with xeroderma pigmentosum have an inherited defect in NER and have a 1000-fold increased risk of skin cancer.

- Many patients with hereditary non-polyposis colorectal cancer (HNPCC) have an inherited defect in mismatch repair.

- The major types of chemotherapies are:
  - Alkylating agents. Two examples are: clorambucil and cisplatin.
  - Antimetabolites. Two examples are: 5FU and methotrexate.
  - Organic drugs: Two examples are: vincristine and vinblastine.

- The development of drug resistance is a major problem for chemotherapy.

■ ACTIVITY

Make a list of five carcinogens and the mutations they cause. Describe the method of DNA repair used to correct each type of mutation.

■ FURTHER READING

Chabner, B.A. and Longo, D.L. (2001) *Cancer chemotherapy and biotherapy—principles and practice*, 3rd edn. Lippincott Williams and Wilkins, Philadelphia.

Gottesman, M.M. (2002) Mechanisms of cancer drug resistance. *Annu. Rev. Med.* **53**:615–627.

Hecht, S.S. (2003) Tobacco carcinogens, their biomarkers and tobacco-induced cancer. *Nature Rev. Cancer* **3**:733–737.

Hoeijmakers, J.H.J. (2001) Genome maintenance mechanisms for preventing cancer. *Nature* **411**:366–374.

Ichihashi, M., Ueda, M., Budiyanto, A., Bito, T., Oka, M., Fukunaga, M., Tsuru, K. and Horikawa, T. (2003) UV-induced skin damage. *Toxicology* **189**:21–39.

Lane, N. (2002) Oxygen—The molecule that made the world. Oxford University Press, Oxford.

Lax, A.J. and Thomas, W. (2002) How bacteria could cause cancer: one step at a time. *Trends Microbiol.* **10**:293–299.

Pfeifer, G.P., You, Y.-H. and Besaratinia, A. (2004) Mutations induced by ultraviolet light. *Mut. Res.* in press.

Williams, G.M. and Jeffrey, A.M. (2000) Oxidative DNA damage: endogenous and chemically induced. *Regulatory Toxicol. Pharmacol.* **32**:283–292.

■ SELECTED SPECIAL TOPICS

Alexandrov, K., Cascorbi, I., Rojas, M., Bouvier, G., Kriek, E. and Bartsch, H. (2002) CYP1A1 and GSTM1 genotypes affect benzo[a] pyrene DNA adducts in smokers' lung: comparison with aromatic/hydrophobic adduct formation. *Carcinogenesis* **23**:1969–1977.

Al-Tassan, N., Chmiel, N.H., Maynard, J., Fleming, N., Livingston, A.L., Williams, G.T., Hodges, A.K., Davies, D.R., David, S.S., Sampson, J.R. *et al.* (2002) Inherited variants of MYH associated with somatic G:C → T:A mutations in colorectal tumors. *Nat. Genet.* **30**:227–232.

Davies, H., Bignell, G.R., Cox, C., Stephens, P. *et al.* (2002) Mutations of the BRAF gene in human cancer. *Nature* **417**:949–954.

Kelley, J.R. and Duggan, J.M. (2003) Gastric cancer epidemiology and risk factors. *J. Clin. Epidemiol.* **56**:1–9.

Preston, D.L., Shimizu, Y., Pierce, D.A., Suyama, A. and Mabuchi, K. (2003) Studies of mortality of atomic bomb survivors. Report 13: solid cancer and noncancer disease mortality; 1950–1997. *Radiation Res.* **160**:381–407.

Smith, C.J., Perfetti, T.A., Garg, R. and Hansch, C. (2003) IARC carcinogens reported in cigarette mainstream smoke and their calculated log P values. *Food Chem. Toxicol.* **41**:807–817.

Whitmore, S.E., Morison, W.L., Potten, C.S. and Chadwick, C. (2001) Tanning salon exposure and molecular alterations. *J. Am. Acad. Dermatol.* **44**:775–780.

3

Regulation of gene expression

Introduction

Cancer is a genetic disease at the cellular level which may be manifested by alterations in gene expression. In this chapter we will review the molecular components involved in transcriptional regulation. As mentioned in Chapter 2, mutations in the promoter region of genes can alter the regulation of gene expression and lead to carcinogenesis. More recently, an additional mechanism of regulating gene expression called epigenetics (Greek for "upon" the genome) has been proposed to be important for carcinogenesis. It involves alterations in inheritable information encoded by modifications of the genome and chromatin components. The structure of a gene within the context of chromatin is described below in order to elucidate how gene and chromatin structure affects gene expression. Note also that in this chapter there is a focus on DNA–protein interactions in transcriptional regulation, chromatin configuration, and telomere extension.

3.1 Transcription factors and transcriptional regulation

Transcription factors are proteins that bind to gene promoters and regulate transcription. They contain a set of independent protein modules or domains, each having a specific role important for transcription factor function. They include DNA binding domains, transcriptional activation domains, dimerization domains, and ligand binding domains.

to power the appliance. A DNA binding domain is the part of the transcription factor whose function is to recognize specific DNA promoter sequences and bind DNA. There is some variety in the structure of a domain in the same way that there are American, UK, and Continental Europe plugs.

Four common types of DNA binding domains are the helix-turn-helix motif, the leucine zipper motif, the helix-loop-helix motif, and the zinc finger motif. These domains are characteristic protein conformations that enable a transcription factor to bind DNA. It is the conformation of these protein domains that facilitate binding to DNA. Take the helix-turn-helix and zinc finger domains as examples. The amino acid side-chains of the alpha helix lie in the major groove of the DNA helix and hydrogen bond to specific DNA base pairs. The zinc finger domain (approximately 30 amino acids long; Figure 3.1a) is configurated around a zinc atom that links two cysteines and two histidines (shown in red) (or two cysteines and two cysteines). It consists of a simple ββα fold (Figure 3.1b). The side chains of specific amino acids recognize a specific DNA sequence (about five nucleotide pairs). Transactivation domains function

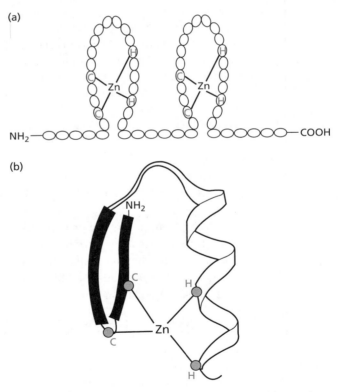

Figure 3.1 The zinc finger DNA binding domain: Primary (a) and secondary (b) structure

by binding to other components of the transcriptional apparatus in order to induce transcription by RNA polymerase, the main enzyme required for transcription. Some transcription factors work in pairs or dimers and require a dimerization domain which facilitates protein–protein interactions between them. Interactions between transcription factors are a common theme in transcriptional regulation. Some transcription factors only function upon binding of a ligand and therefore require a ligand binding domain (this is analogous to the space for a coin in a pinball machine). The activity of a transcription factor can be regulated by several means: synthesis in particular cell types only, covalent modification such as phosphorylation, ligand binding, cell localization, and/or if dimeric, by exchange of partner proteins.

Many of the key points of transcriptional regulation can be demonstrated by two examples: the AP-1 transcription factor family and the steroid hormone receptors.

The AP-1 transcription factor is important for the processes of growth, differentiation, and death and therefore plays a role in carcinogenesis. AP-1 binds either to the 12-O-tetradecanoylphorbol-13-acetate (TPA) response element or the cAMP response element in the promoter region of their target genes. The transcription factor is actually composed of two components and can be produced by dimers of proteins from the Jun and Fos families (Jun, Jun B, Jun D, Fos, Fos B, FRA1, and FRA 2) (Figure 3.2). Eighteen possible combinations are possible. Both Jun and Fos members contain a basic leucine zipper dimerization domain. Because the processes of growth, differentiation, and apoptosis need to be

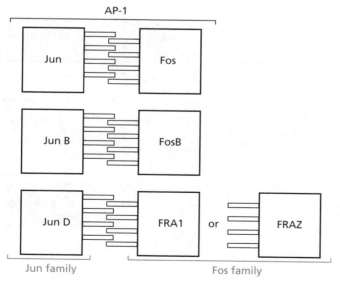

Figure 3.2 Members of the Jun and Fos transcription factor family

carefully regulated, AP-1 is itself activated in response to specific signals such as growth factors, ROS, and radiation. The specific combination of the dimer composition influences the biological response. The antagonism displayed between Jun and Jun B, with regard to cell proliferation in some cell types, supports this: Jun acts as a positive regulator of proliferation while Jun B acts as a negative regulator in the presence of Jun.

The family of steroid hormone receptors acts as ligand dependent transcription factors. They contain a zinc finger type of DNA binding domain, a ligand binding domain for a specific steroid hormone, and a dimerization domain since they activate transcription as a dimer. Each domain functions independently and in a manner that is specific for a particular steroid hormone receptor. This feature has been utilized as a molecular tool by scientists in so-called domain swap experiments which produce chimaeric receptors. For example if the ligand binding domain of the thyroid hormone receptor is swapped with the ligand binding domain of the retinoic acid receptor, the newly formed chimaeric receptor (Figure 3.3) will retain the DNA binding domain of the thyroid hormone receptor and will activate thyroid hormone responsive genes. However, these genes will be activated by retinoic acid via the retinoic acid ligand binding domain, and not by thyroid hormone. Such experiments clearly demonstrate the functional independence of these domains.

Steroid hormones pass through the cell membrane and bind to their particular intracellular receptors. Upon binding, the receptors move into the nucleus and activate transcription of their target genes through specific response elements. The retinoic acid receptor (RAR), as a member of

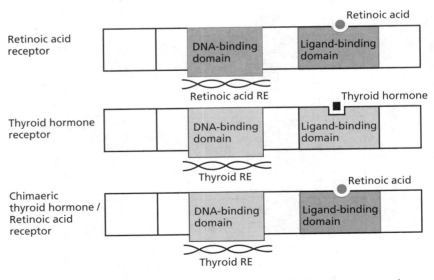

Figure 3.3 A chimaeric steroid hormone receptor is shown below its two parental receptors

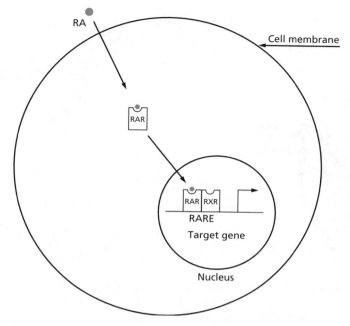

Figure 3.4 Mechanism of action of the retinoic acid receptor: RAR

the steroid hormone receptor family, acts as a retinoic acid (RA) dependent transcriptional regulator and is important during differentiation. It binds to the RA response element (RARE) in target genes as a heterodimer with another member of the family called RXR (Figure 3.4). Aberrant forms of RARs are characteristic of several leukemias. As we will see below, steroid hormone receptors play an important role in many different types of cancer.

3.2 Chromatin structure

Human DNA is present in the nucleus of cells in the form of 46 chromosomes. Chromosomes are made of chromatin: a thread of DNA (60%) plus associated RNA (5%) and protein (35%). It is astonishing to think that the actual length of DNA in the nucleus of a cell is over a meter when fully extended. A high level of packaging (Figure 3.5) is required neatly to organize the DNA to fit into the nucleus of a cell and to allow it to assume necessarily organized conformations for transcription and replication. Both of these processes involve denaturation of the double helix and reading of template strands.

The simplest or primary level of organization of chromatin is the wrapping of DNA around a protein "spool" and is referred to as the

PAUSE AND THINK

How would you organize long pieces of thread in a sewing box? Most thread is wrapped around a spool for orderly and easy unwinding. Few people would just form a disorganized bunch of thread for fear of getting tangles and knots.

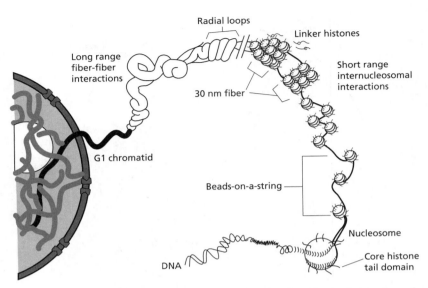

Figure 3.5 Multiple levels of chromatin structure. Reprinted, with permission, from the *Ann. Rev. Biophys. Biomolec. Struct.*, Vol. 3, p. 362, Copyright 2002 by Annual Reviews www.annualreviews.org

"beads on a string" array. The beads represent the nucleosome, which contains 147 base pairs (bp) of DNA wrapped 1.7 times around a core of histone proteins. The histone core is an octomer of histones containing two copies of H2A, H2B, H3, and H4 histones. Each histone contains domains for histone–histone and histone–DNA interactions and NH2-terminal lysine-rich and COOH-terminal "tail" domains which can be post-translationally modified (e.g. acetylated, methylated, or phosphorylated). Histone H1 is a linker histone and binds to DNA located outside the core. Ten to 60 base pairs of DNA separate the "beads". The secondary level of organization is the formation of 30 nm fibers and these can associate to form tertiary structure radial loops.

Chromatin has an important role beyond being a structural scaffold. The degree of compaction or relaxation of chromatin structure can change and it is this feature that enables it to have a regulatory role in transcription. Information for the memory and inheritance of chromatin conformation is encoded by epigenetic modifications.

3.3 Epigenetic regulation of transcription

Epigenetics refers to inheritable information that is encoded by modifications of the genome and chromatin components. These modifications affect the structure and conformation of chromatin and, consequently, transcriptional regulation. Epigenetic alterations in gene expression do not cause a change in the DNA nucleotide sequence and therefore are not mutations. Stable epigenetic switches are important during normal cell differentiation. (Note: it is differential gene expression that makes one cell type different from another.) Two types of epigenetic mechanisms will be discussed below: histone modifications and DNA methylation. Both can be acquired or inherited and both affect transcriptional activity by regulating access of transcription factors to appropriate nucleotide sequences in gene promoters.

Histone modification

Histone proteins are subject to diverse post-translational modifications such as acetylation, methylation, phosphorylation, and ubiquination. The histone code hypothesis predicts that the pattern of these multiple histone modifications specify the components and activity of the transcription regulatory molecular machinery. Let us focus on acetylation.

The acetylation pattern of histones alters chromatin structure and affects gene expression (Figure 3.6). Acetylation acts as a docking signal for the recruitment or the repulsion of chromatin-modifying factors. Histone acetyltransferases (HATs; add acetyl groups) and histone

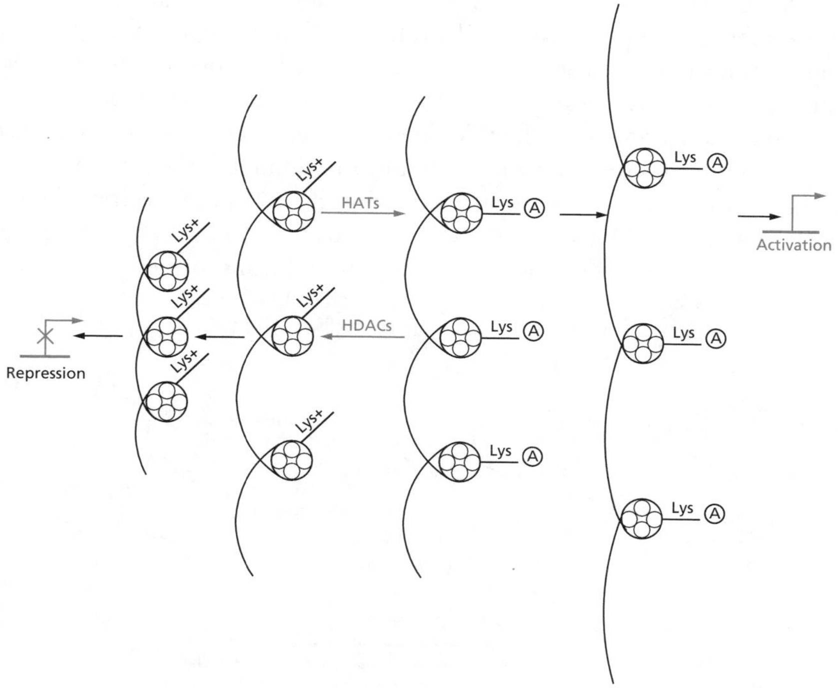

Figure 3.6 Histone acetylation affects gene expression

deacetylases (HDACs; remove acetyl groups) are two families of enzymes that produce the pattern. HATs acetylate specific histone-tail lysines and other non-histone proteins, including transcription factors (e.g. E2F and p53). HATs relax chromatin folding by regulating the binding of non-histone proteins and this correlates with enhanced transcriptional elongation by RNA polymerase III. Histone deacetylases (HDACs) remove acetyl groups and restore a positive charge to lysine residues of the histone tails which stabilize chromatin compaction and higher level packaging. This configuration of chromatin limits the accessibility of transcription factors and results in the repression of transcription.

In addition, transcriptional activators often recruit HATs, and other chromatin-remodeling enzymes to the promoter region. The retinoblastoma tumor suppressor protein mentioned as a molecule of fame in Chapter 1 exerts its effects, in part, by recruiting HDACs to specific gene promoters (see Chapter 5). Thus, a signaling network seems to underlie chromatin modeling.

PAUSE AND THINK

In general, HATs activate transcription and HDACs repress transcription.

DNA methylation

Another epigenetic process that affects transcriptional regulation is DNA methylation. DNA methylation is the addition of a methyl group to position 5 of cytosine. Only 3–4% of all cytosines are methylated. Methylation only occurs at cytosine nucleotides which are situated 5′ to guanine nucleotides (CpGs). Methylcytosine deaminates spontaneously and results in C → T transitions (Figure 3.7). It is thought that evolution has selected against this dinucleotide due to the high rate of mutation, since CpG is under-represented and unequally distributed in the genome. CpG clusters, called CpG islands, are located in the promoter region of 50% of human genes. In general, the CpG islands found in gene promoter regions are not methylated in normal tissues and transcription may occur.

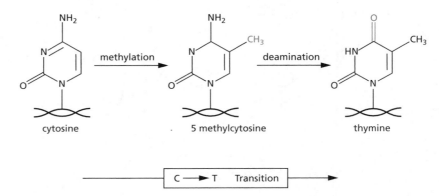

Figure 3.7 Spontaneous deamination of methylcytosine leads to a C → T transition

Methylated cytosines are found mainly in repetitive sequences and in the CpG islands found in the promoter region of repressed genes such as X-chromosome inactivated genes, imprinted genes, and some tissue-specific genes. In these cases, methylation is a heritable signal that is associated with a compacted chromatin structure and maintains gene silencing.

Enzymes called DNA methyltransferases (DNMTs) mediate the covalent addition of a methyl group from the methyl carrier S-adenosylmethionine cytosine. Three methyltransferases are known: DNMT1, DNMT3a, DNMT3b. DNMT1 is involved in the conversion of hemimethylated DNA to fully methylated DNA during replication. This mechanism allows methylation patterns to be inheritable; if only one strand remained methylated the signal would be lost in half of its daughter cells after replication. The other two methyltransferases are mainly involved in *de novo* methyltransferase activity (methylation of new sites).

It has been suggested that the mechanism by which methylation results in silencing is by recruiting methyl binding domain (MBD) proteins, which have been shown to interact with HDACs and chromatin remodeling enzymes. Therefore, epigenetic regulation of transcription includes cross-talk between methylation, chromatin-remodeling enzymes and histone modification.

3.4 Evidence for a role of epigenetics in carcinogenesis

The fairly recent and still controversial proposal that epigenetic inactivation of genes is as important as the inactivation of genes by mutation during carcinogenesis has recently been put forth. A brief examination of some of the accumulating supporting evidence is described below.

Histone modification and cancer

Altered HAT or HDAC activity has been observed in several cancers. Interestingly, one gene, EP300, that codes for a HAT has been found to be mutated in epithelial cancers. Several of the mutations predicted a truncated protein and inactivation of the second allele was observed in 5/6 cases, suggesting that it functions as a tumor suppressor (Gayther *et al.*, 2000).

Acute promyelocytic leukemia is characterized by a chromosomal translocation that produces a fusion protein called PML-RAR. This novel fusion protein retains the DNA binding domain and ligand binding domain of the RAR in addition to PML sequences. PML-RAR recruits HDAC to the promoter region of RA target genes and represses the expression of these genes. The lack of activation of RAR target genes causes the block of differentiation that characterizes the leukemia. Other tumors also have been associated with aberrant recruitment of HDACs.

PAUSE AND THINK

RAR is a member of which family of transcription factors? See Section 3.1.

Methylation and cancer

Cancer-specific changes in DNA methylation have been recognized, and many studies have focused on hypermethylation observed in normally unmethylated CpG islands of gene promoters (see Box below). Gene silencing by methylation may be an important mechanism of carcinogenesis whereby critical genes normally involved in tumor suppression may be switched off. Inactivation of gene expression by methylation of promoter regions of such genes has been observed in cancer cell lines and human tumors. For example, estrogen receptor protein is present in normal ovarian epithelial cells but is frequently lost in ovarian cancer. Hypermethylation of the estrogen receptor-α gene promoter was observed in three out of four human ovarian cell lines that lacked estrogen receptor protein (O'Doherty *et al.*, 2002). This indicates that hypermethylation may be responsible for this phenotype in ovarian tumors. As another example, the breast cancer susceptibility gene, *BRCA1*, is often mutated in a recessive manner in inherited breast cancer. Thus, the loss of function of the gene product suggests that normal BRCA1 acts to suppress breast cancer. Mutation of *BRCA1* is very rarely observed in non-inherited breast cancer. However, interestingly hypermethylation is associated with the inactivation of *BRCA1* in non-inherited breast cancer and therefore this may be another way of accomplishing loss of function. These findings support the view that epigenetics may be an additional mechanism for carcinogenesis. Additional examples of some key target genes affected by methylation include the *retinoblastoma (Rb)* gene; inhibitor of the cell cycle *p16 INK4a*; pro-apoptotic death-associated protein kinase (*DAPK*); *APC*; and the *estrogen receptor* gene.

Analysis of DNA methylation by sodium bisulfate treatment and methylation-specific PCR

Molecular biology procedures used for standard genetic analysis erase DNA methylation information and so specialized methods for methylation analysis were developed. Sodium bisulfate treatment of genomic DNA converts unmethylated cytosine residues to uracil by deamination. 5-methylcytosines are converted to thymine under these same conditions. Treated DNA is no longer complementary and PCR amplification requires specially designed primers. There are several possible designs for PCR amplification primers of the resulting DNA but, most commonly, primers are designed to hybridize specifically with the sodium bisulfate modified sequences. PCR that uses these types of primers is called methylation-specific PCR. Methylation-specific PCR provides information about particular methylation patterns. Analysis of many sites throughout the genome can be collected to produce a methylation profile.

The molecular mechanisms underlying specific methylation events are largely unknown. DNA methylation by itself does not directly repress transcription but requires associated proteins such as histone-modifying enzymes (described above), methyl cytosine binding proteins, and DNMTs. These proteins collaborate to define the structure of chromatin. In addition to recruiting HDAC (discussed above), PML-RAR has also been shown to recruit methyltransferase resulting in the DNA methylation of a promoter region of a specific gene (Di Croce *et al.*, 2002). The RARβ2 gene has a RARE in its promoter and is one of the target genes of PML-RAR. It has been demonstrated that PML-RAR forms stable complexes with DNMTs at the RARβ2 promoter and that the resulting hypermethylation contributes to carcinogenesis. It has been suggested that DNMTs, in addition to mediating methylation, may act as a platform for the assembly of chromatin-modifying factors (see Pause and Think).

Since not all carcinogens are mutagens, it may be possible that some non-genotoxic carcinogens (agents that do not mutate genes) are epigenetic carcinogens. Hypermethylation has been observed in tumor-sensitive mice treated with phenobarbital, a non-genotoxic carcinogen in rodents (Watson and Goodman, 2002). The data suggest that a disruption to normal methylation patterns is related to tumor susceptibility and that the mechanism of non-genotoxic carcinogens may act via methylation. In addition, nutritional deficiencies (methionine, choline) seem to affect the cellular level of S-adenosylmethionine, an important methyl group donor. This suggests that perturbation of methylation can be produced through the diet (see Chapter 9). Additional direct evidence is needed to link factors that are necessary for inducing the misregulation of methylation and for inducing carcinogenesis. Or perhaps mutation is really the only culprit responsible for the misregulation of methylation? Mutation of DNA methyltransferases demonstrated in several cancers would lead directly to altered methylation, but this does not account for the specificity of tumor suppressor genes. Inversely, the role of methylation in transformation may be to promote mutation. Methylated cytosine residues have a tendency to deaminate spontaneously causing C to T transitions. This may account for the increased mutation rate observed in methylated CpG islands.

Paradoxically, the genome of a cancer cell overall can have 20–60% less methylation than a normal cell. This global hypomethylation, mainly in the coding region of genes and of repetitive DNA sequences, occurs in the cancer cell at the same time as the hypermethylation of specific genes described above. This results in the activation of genes not normally expressed. Although this phenomenon has not been vigorously studied, one report points to a causal role of DNA hypomethylation in tumor formation (Gaudet *et al.*, 2003). Mice generated to have reduced levels of DNA methyltransferase 1 exhibited genome-wide hypomethylation and

PAUSE AND THINK

The data examined above, supporting the hypothesis that epigenetics is as important as mutation for carcinogenesis, is lacking a critical piece of evidence: has a particular carcinogen caused the alterations in epigenetic regulation? The evidence that carcinogens cause mutations is indisputable.

developed T-cell lymphomas. Overall, epigenetics and cancer is an area where further research is obviously needed.

3.5 Telomeres and telomerase

One of the hallmarks of cancer cells is that they acquire a limitless replicative potential (see Figure 1.1). Normal cells have an autonomous program that allows for a finite number of replication cycles. This phenomenon is well known *in vitro* in that cells in culture only undergo a certain number of doublings before they stop dividing and enter senescence (permanent growth arrest). Telomeres, repetitive DNA sequences at the ends of chromosomes, have been shown to function as a molecular counter of the cell's replicative potential. Telomeres protect the ends of chromosomes and are composed of several thousand repeats of the sequence TTAGGG bound by a set of specific proteins. Telomeres shorten by 50–200 bp with each round of DNA replication due to the limits of DNA polymerases during DNA replication (see Box: 'A little lesson about DNA replication' below). These enzymes proceed only in the 5–3′ direction and require an RNA primer to initiate DNA synthesis. The RNA primers are removed after replication is complete. As a result, the 3′-end of the parental chromosomal DNA is not replicated and thus chromosomes progressively erode during each round of replication (Figure 3.8). When the chromosomes reach a threshold length, cells enter a stable state of growth arrest called cellular senescence. If cells bypass this stage due to mutation and telomeres become critically short, chromosomal instability results and apoptosis (or cell transformation discussed below) is induced. Maintaining telomere length in stem cells of renewal tissue (e.g. the basal layer of the epidermis) is important for providing a longer replicative potential. Telomerase, a ribonucleoprotein containing human telomerase reverse transcriptase activity (hTERT) and a human telomerase RNA (hTR) maintain telomere length in certain cell types, such as stem cells. Reverse transcriptases are enzymes that synthesize DNA from RNA—the reverse of the central dogma of molecular biology. The hTR contains 11 complementary base pairs to the TTAGGG

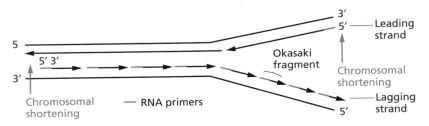

Figure 3.8 Chromosomal shortening after DNA replication

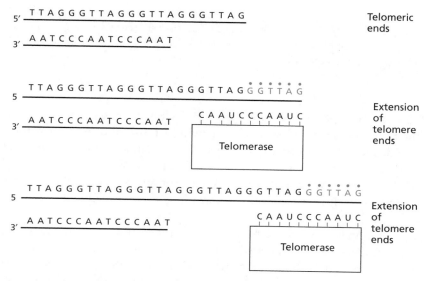

Figure 3.9 Telomere extension by telomerase

repeats and acts as a template for the reverse transcriptase to add new repeats (shown in red) to telomeric DNA on the 3′-ends of chromosomes (Figure 3.9).

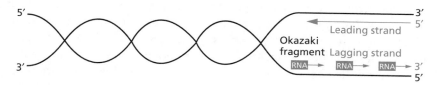

Figure 3.10 Semiconservative and semidiscontinuous DNA replication

A little lesson about DNA replication...

DNA replication proceeds in a semi-conservative manner: Each of the two parental strands acts as a template for the synthesis of a newly replicated strand (Figure 3.10; new DNA strand synthesis is shown in red). Each of the polynucleotide strands that make up the DNA helix has a sense of direction; that is each has a 5′-end and a 3′-end. The two strands are arranged in an antiparallel manner. Since DNA polymerases only work in a 5′–3′ direction each strand is replicated differently as the DNA helix unwinds. For one strand, the leading strand, replication proceeds in a continuous manner from the 5′–3′ end. For the other strand, the lagging strand, replication occurs in a discontinuous manner through the 5′–3′ synthesis of short Okazaki fragments. After removing the RNA primers and filling in the gaps, these fragments are ligated together by →

→ the enzyme DNA ligase to form one continuous strand. The requirement for a RNA primer by DNA polymerase and the subsequent removal of this primer cause the strands to shorten at the extreme chromosomal ends during each round of replication.

It has been shown *in vitro* that the telomere ends are not linear but rather complicated structures forming t-loops and may form four-stranded DNA conformations called G quadruplexes. It is important that chromosome ends are distinguishable from DNA double-stranded breaks. If they were not, the DNA repair processes would produce chromosomal fusions and other aberrations in an attempt to repair the damage.

Several lines of evidence have linked telomerase activity with cancer. The maintenance of telomeres seems to be important for tumor growth and approximately 90% of tumors accomplish this by upregulating telomerase. Telomerase activity was clearly a distinguishing feature in one classical study where it was detectable in cultured immortal cell lines (98 of 100) and tumor tissue biopsies (90 of 101) but undetectable in cultured normal somatic cells (22) or benign tissue samples (50). It has been found that telomerase, in addition to two oncogenes, is essential in the protocol to transform normal fibroblasts to cancer cells *in vitro*, thus providing a strong link between telomerase and tumorigenesis. Several oncogenes have been demonstrated to regulate the expression of telomerase. For example, the transcription factor c-myc (an oncogene discussed in later chapters) increases the expression of the *hTERT* gene via specific response elements in the promoter region. As mentioned above, if cells bypass the replicative senescence stage due to mutation, telomeres become critically short and chromosomal instability results. This genetic catastrophe may lead to the loss of tumor suppression mechanisms, bypassing the apoptotic trigger and initiating carcinogenesis.

Interestingly, modifications of the telomere hypothesis of senescence described above have recently been suggested and have strong implications for cancer. The telomere hypothesis would predict that telomeres shorten at a constant rate, yet great heterogeneity of replicative lifespan exists among cells within a clonally derived population (i.e. some cells arrest after a few divisions and some after many divisions). It has been reported that telomere shortening is accelerated by oxidative stress (von Zglinicki, 2002), which suggests that the problem of replicating the ends of chromosomes is not the only determining factor for telomere length and replicative potential. Telomeric DNA is repaired less proficiently compared with the bulk of the genome in response to oxidative damage. Unrepaired single-strand breaks accelerate telomere shortening, although the mechanism by which this occurs is unclear. These observations suggest that telomeric DNA may act as a sensor for DNA damage and may explain why there is great heterogeneity in the rate of telomeric shortening among individual cells. Therefore, telomere shortening may act as a tumor suppression mechanism by limiting replicative potential in response to genome damage.

◎ Therapeutic strategies

3.6 Epigenomic and histonomic drugs

It is a fairly recent view that epigenetic silencing may be as important as mutation as a mechanism for carcinogenesis. Currently, the concept of reversing somatic mutations is difficult to envisage. However, the concept may be conceivable for epigenetic changes since these are modifications that are potentially reversible. A large number of genes known to play important roles in carcinogenesis have been shown to display hypermethylation of their promoter regions. Since the increased methylation seen in tumor cells is not observed in normal cells, it provides a tumor-specific target for DNA methylation inhibitors. Similarly, enzymes that alter chromatin structure, such as HDACs that modify histones, provide other molecular targets. As described above, several forms of leukemia and lymphoma are associated with transcriptional repression due to recruitment of HDACs. Reversal of epigenetic silencing is an approach that may lead to new therapeutics.

DNA methylation inhibitors

Drugs that block DNA methylation are predicted to show anti-tumor effects since inactivation of tumor suppressor genes by methylation may be an important mechanism in carcinogenesis. Recall that DNA methylation occurs at position 5 on cytosine. Two 5'-modified analogs of deoxycytidine, 5-azacytidine (5-azaC) and 5-aza-2'-deoxycytidine, have been used to target DNA methyltransferases (Figure 3.11). These drugs are incorporated into DNA and/or RNA. They covalently link with DNA methyltransferases (DNMT, left red target in Figure 3.11) and sequester its action such that there is significant demethylation after several rounds of replication. These drugs may result in DNA instability that parallels antimetabolite chemotherapeutic agents. Another potential hindrance is that aberrant methylation and gene repression return after treatment is stopped, dictating that drug administration must be prolonged. Both of these drugs showed antileukemic activity in clinical trials but were not successful in solid tumors.

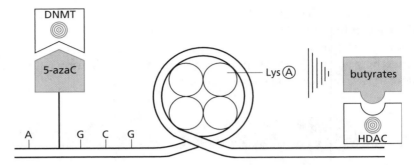

Figure 3.11 Drugs designed to target epigenetic mechanisms (shown in red)

Inhibitors of histone deacetylases

Histone-modifying enzymes have been targeted for the development of new cancer drugs (Figure 3.11). Recall that HDACs generally repress gene transcription and that aberrant recruitment is characteristic of some cancers such as leukemias. Re-activation of silenced genes involved in growth, differentiation, or apoptosis, provides the rationale for treating such cancers with inhibitors of HDACs. Several drugs that bind to the catalytic site of HDACs (right red target in Figure 3.11) and block the binding to their substrates (acetylated lysines of histone proteins) are being tested in clinical trials: butyrates, valproic acid, the hydroxamic acid-based compounds SAHA and pyroxamide, and depsipeptide (FR901228). In general, these drugs are well tolerated and many can be administered orally. Alteration of gene expression appears to be selective. Interestingly, many HDAC inhibitors induce p21^{WAF1}, a cyclin dependent kinase inhibitor important for growth arrest. These drugs seem to have little or no effect on normal cells. Molecular effectiveness was demonstrated by the detection of acetylated histones in particular white blood cells and tumor cells and this was associated with clinical improvement. Butyrates have been approved for use in the clinic.

3.7 Telomerase inhibitors

The relatively tumor-specific expression of telomerase and its pivotal role in the ability of a cancer cell to divide indefinitely suggest that it may be a valuable molecular target for new cancer therapies. However, several parameters need to be examined when considering the inhibition of telomerase as a cancer therapy. Effectiveness may depend on initial telomere length and thus this should be assessed from tumor biopsies prior to treatment. Also the response may be slow due to the time needed for the telomeres of cancer cells to shorten enough to trigger senescence or apoptosis and long-term treatment may be necessary. In general, long-term treatment increases the probability of drug resistance. Several different strategies, targeting either the RNA component or the catalytic protein component, have been explored in preclinical studies. Since telomerase is dependent on its functional RNA molecule, antisense oligonucleotides and ribozymes have been popular agents used to target hTR. Antisense oligonucleotides are complementary to part of the target RNA and hybridize by Watson–Crick base pairing. Hybridization can inhibit function directly or trigger degradation by the recruitment of RNAses. Hammerhead ribozymes contain antisense sequences for target recognition and an internal endonuclease activity that cleaves the target RNA. Reverse transcriptase inhibitors against the catalytic domain hTERT and nucleoside analogs have also been investigated. BIBR1532 is a synthetic small molecule inhibitor that directly binds hTERT non-competitively and has been shown to induce telomere-driven senescence *in vitro* and *in vivo*. G-quadruplex binding molecules that prevent the interaction between the enzyme and substrate have also been developed (e.g. telomestatin). High throughput screening has identified several natural compounds as telomerase inhibitors, such as components of mistletoe and a green tea catechin. Careful clinical trials are needed to see which, if any, anti-telomerase therapies are effective.

■ CHAPTER HIGHLIGHTS—REFRESH YOUR MEMORY

- In simplistic terms, a gene consists of a regulatory region and a coding region. Mutations in the former may alter gene expression while mutations in the latter may affect the gene product.

- Transcription factors recognize DNA response elements and are essential for the regulation of gene expression.

- Steroid hormone receptors act as ligand-dependent transcription factors.

- Chromatin has several levels of DNA packaging: the nucleosome, the 30 nm fiber, and radial loops.

- Epigenetic changes also regulate gene expression. These involve modification of nucleotides or chromatin components.

- Histone modification and methylation are two types of epigenetic mechanisms.

- HATs add acetyl groups to histones and activate transcription.

- DHACs remove acetyl groups and repress transcription.

- Methylation at CpG islands repress transcription.

- Evidence is accumulating for the role of epigenetic inactivation in carcinogenesis.

- Telomeres play a role in the replicative potential of a cell.

- Telomeres shorten with each round of replication but the rate of shortening may also be influenced by oxidative stress.

- Telomerase is an enzyme that maintains telomere length.

- Telomerase activity is increased in 90% of tumors.

- Strategies for the design of new drugs target DNMTs, HDACs, and telomerase.

■ **ACTIVITY**

(i) Formulate evidence for your view on the statement that epigenetics is as important as mutation for carcinogenesis. Contribute to a class debate on this issue.

■ **FURTHER READING**

Baylin, S.B. and Herman, J.G. (2000) DNA hypermethylation in tumorigenesis: epigenetics joins genetics. *Trends Genet.* **16**:168–174.

Brown, R. and Strathdee, G. (2002) Epigenomics and epigenetic therapy of cancer. *Trends Mol. Med.* **8**:(Suppl.) S43–S48.

Esteller, M. and Herman, J.G. (2002) Cancer as an epigenetic disease: DNA methylation and chromatin alterations in human tumors. *J. Pathol.* **196**:1–7.

Goffin, J. and Eisenhauer, E. (2002) DNA methyltransferase inhibitors—state of the art. *Ann. Oncol.* **13**:1699–1716.

Herman, J.G. and Baylin, S.B. (2003) Gene silencing in cancer in association with promoter hypermethylation. *N. Eng. J. Med.* **349**:2042–2054.

Kelland, L.R. (2001) Telomerase: biology and phase 1 trials. *The Lancet Oncol.* **2**:95–102.

Laird, P.W. (2003) The power and the promise of DNA methylation markers. *Nature* **3**:253–266.

Marks, P.A., Richon, V.M., Breslow, R. and Rifkind, R.A. (2001) Histone deacetylase inhibitors as new cancer drugs. *Curr. Opin. Oncol.* **13**:477–483.

Rezler, E.M., Bearss, D.J. and Hurley, L.H. (2002) Telomeres and telomerases as drug targets. *Current Opin. Pharm.* **2**:415–423.

Schreiber, S.L. and Bernstein, B.E. (2002) Signaling network model of chromatin. *Cell* **111**:771–778.

Shaulian, E. and Karin, M. (2002) AP-1 as a regulator of cell life and death. *Nature Cell Biol.* **4**:E131–E136.

White, L.K., Wright, W.E. and Shay, J.W. (2001) Telomerase inhibitors. *Trends Biotech.* **19**:114–120.

■ **WEB SITES**

DNMT1 antisense http://www.methylgene.com

■ **SELECTED SPECIAL TOPICS**

Di Croce, L., Raker, V.A., Corsaro, M., Fazi, F., Fanelli, M., Faretta, M., Fuks, F., Lo CoCo, F., Kouzarides, T., Nervi, C., Minucci, S. and Pelicci, P.G. (2002) Methyltransferase recruitment and DNA hypermethylation of target promoters by an oncogenic transcription factor. *Science* **295**:1079–1082.

Gaudet, F., Hodgson, J.G., Eden, A., Jackson-Grusby, L., Dausman, J., Gray, J.W., Leonhardt, H. and Jaenisch, R. (2003) Induction of tumors in mice by genomic hypomethylation. *Science* **300**:489–492.

Gayther, S.A., Batley, S.J., Linger, L., Bannister, A., Thorpe, K., Chin, S.-F., Daigo, Y., Russell, P., Wilson, A., Sowter, H.M., Delhanty, J.D.A., Ponder, B.A.J, Kouzarides, T. and Caldas, C. (2000). Mutations truncating the EP300 acetylase in human cancers. *Nature Gen.* **24**:300–303.

O'Doherty, A.M., Church, S.W., Russell, S.E.H., Nelson, J. and Hickey, I. (2002) Methylation status of oestrogen receptor-a gene promoter sequences in human ovarian epithelial cell lines. *Br. J. Cancer* **86**:282–284.

von Zglinicki, T. (2002) Oxidative stress shortens telomeres. *Trends Biochem. Sci.* **27**:339–344.

Watson, R.E. and Goodman, J.I. (2002) Effects of phenobarbital on DNA methylation in GC-rich regions of hepatic DNA from mice that exhibit different levels of susceptibility to liver tumorigenesis. *Toxicolog. Sci.* **68**:51–58.

4

Growth factor signaling and oncogenes

Introduction

One of the fundamental characteristics of cells is their ability to self-reproduce. The process of cell division (also known as cell proliferation or cell growth) must be carefully regulated and DNA replication must be precisely coordinated in order to maintain the integrity of the genome for each cell generation. As emphasized earlier in this volume, unregulated growth is a quintessential characteristic of cancer.

An extracellular growth factor stimulates cell growth by transmitting a signal into the cell, and ultimately to the nucleus, to regulate gene expression in order to produce proteins that are essential to cell division. There are four types of proteins involved in the transduction of a growth factor signal: besides growth factors, there are growth factor receptors, intracellular signal transducers, and nuclear transcription factors which elicit the mitogenic effect through the regulation of gene expression. Examining the normal mechanism of growth will allow a better understanding of the abnormalities that occur during carcinogenesis.

It is important to identify a common thread in many growth factor signal transduction pathways: many growth factor receptors are tyrosine kinases. All kinases catalyze the transfer of the γ phosphate group from ATP/GTP to hydroxyl groups on a specific amino in a target protein. Tyrosine kinases phosphorylate tyrosine residues in target proteins (Figure 4.1). Serine/threonine kinases phosphorylate serine and threonine residues. The addition of the phosphate group, a bulky charged molecule, may serve as a recognition site for new protein–protein interactions and/or may cause a conformational change resulting in the activation or inactivation of an enzymatic activity. Specific examples will be described below.

4.1 Epidermal growth factor signaling: an important paradigm

Epidermal growth factor (EGF) and its family of receptor tyrosine kinases serve as an important paradigm for how a signal from an extracellular growth factor can be transduced through a cell, regulate gene

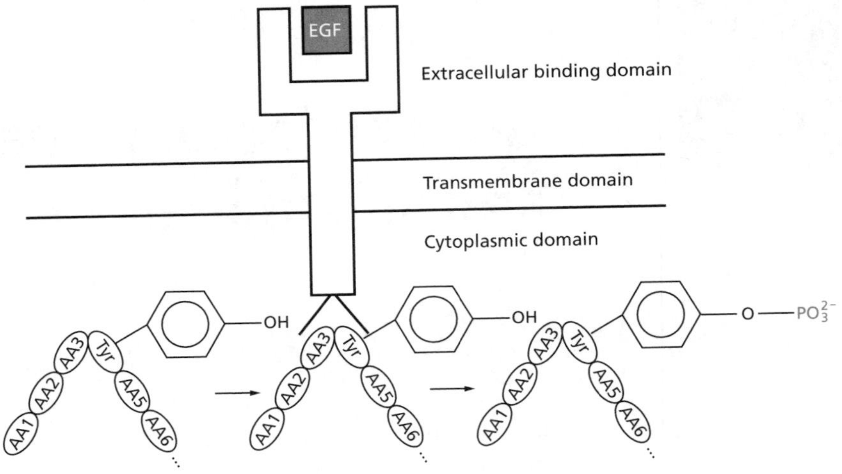

Figure 4.1 Tyrosine kinase receptors phosphorylate tyrosine residues

expression, and trigger cell proliferation. It is a model that we know a great deal about. The EGF receptor (EGFR; also known as ErbB1 or HER1), is a tyrosine kinase receptor and was the first to be discovered. Three additional family members have since been identified: ErbB2 (HER2), ErbB3 (HER3), and ErbB4 (HER4). As members of the receptor tyrosine kinase receptor family, these receptors contain an extracellular ligand binding domain, a single transmembrane domain, and a cytoplasmic protein tyrosine kinase domain (Figure 4.1). Getting the signal from a growth factor outside the cell to inside the nucleus where gene expression is regulated requires several steps: binding of the growth factor to the receptor, receptor dimerization, autophosphorylation, activation of intracellular transducers, including the 'star player' RAS and a cascade of serine/threonine kinases, and regulation of transcription factors for gene expression. Each of these steps involved in the signal transduction pathway of EGF is illustrated in Figure 4.2 and described below. It is essential that you learn this model system because it will enable you to understand many other signal transduction pathways and it will be the basis for illustrating the mechanisms of carcinogenesis. Most interestingly, the components of this pathway have been targets for the design of new cancer therapeutics, some of which will be described at the end of the chapter.

Growth factor binding

The first step in the EGF signal transduction pathway is the binding of EGF to its receptor, EGFR. Extracellular domains (I and III) of EGFR form a binding pocket for the ligand.

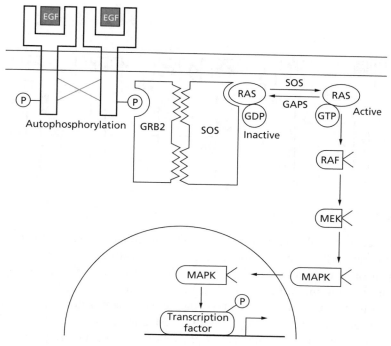

Figure 4.2 The signal transduction pathway of EGF

Dimerization

Dimerization is the process of two EGFR monomers interacting to form a dimer. The mechanism for receptor dimerization as suggested from structural studies is described below and illustrated in Figure 4.3. The binding of one EGF molecule to one receptor causes a conformational change that reveals an extracellular receptor dimerization domain (shown in red). This facilitates the binding to a similar domain in another EGF-bound receptor monomer resulting in a receptor dimer. It is important to note that EGFR can also form heterodimers with other members of the ErbB family. In general, it is assumed that receptors without a ligand are unable to dimerize, although ErbB2 may be an exception and may not have a ligand.

Autophosphorylation

The close proximity of two receptors, facilitated by dimerization, enables the kinase domain of one receptor of the dimer to phosphorylate the other receptor of the dimer and vice versa. This intermolecular (between molecules) autophosphorylation on the cytoplasmic domain of the receptors (shown by a red "X" in Figure 4.2) occurs at tyrosine residues in the activation loop and results in a conformational change. The change

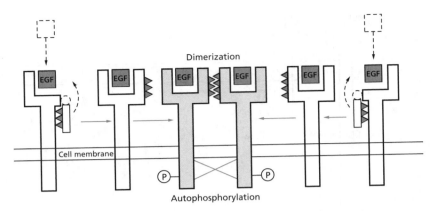

Figure 4.3 EGF receptor dimerization

in receptor conformation permits ATP and substrate access to the catalytic kinase domain. Autophosphorylation is also crucial for the recruitment of cytoplasmic proteins as we will see below. At this stage a signal
from outside the cell has been transduced to inside the cell.

Note that activation of the tyrosine kinase receptor needs to be turned
off after a particular length of time. Mechanisms for termination of
kinase activity include additional phosphorylation triggering a conformational change that inhibits extracellular ligand binding and kinase activity, dephosphorylation of regulatory phosphorylated tyrosine residues by
tyrosine phosphatases, and receptor endocytosis and degradation.

Translocation of specific proteins to the membrane

Activation of the tyrosine kinase catalytic domain facilitates further
phosphorylation. Some phosphorylated tyrosine residues create high affinity binding sites for proteins that contain SH2 (Src homology 2) domains and act as docking sites for the recruitment of specific intracellular proteins.

PAUSE AND THINK

Remember that a domain is a part of a protein with a specific configuration that has
a specific function analogous to an electric plug; see Chapter 3.

SH2 domains (approximately 100 amino acids long) and SH3 (Src homology 3) domains (approximately 50 amino acids long) mediate protein–protein interactions in
pathways activated by tyrosine kinases. SH2 domains recognize and bind to distinct
amino acid sequences (1–6 residues) C-terminal to the phosphorylated tyrosine residue
and SH3 domains recognize and bind to proline and hydrophobic amino acid residues.
Both SH2 and SH3 domains are frequently found in the same protein. Proteins mentioned later in the text that contain SH2 and SH3 domains include SRC, ABL, Grb2,
and PI3-K.

Grb2, an intracellular protein that contains SH2 and SH3 domains, recognizes the phosphorylated receptor via its SH2 domains and facilitates the recruitment of specific proteins to the membrane via its SH3 domains. Specifically, the two SH3 domains of Grb2 interact with the exchange protein SOS (son of sevenless), which facilitates the activation of the pivotal intracellular transducer RAS. Thus, the activator of RAS is translocated from the cytoplasm to the membrane in response to growth factor stimulation.

RAS activation

The RAS proteins are "star players" in regulating cell growth because of their position in the signal transduction pathway; they act as a pivotal point for the integration of a growth factor signal initiating from the membrane with a number of crucial signaling pathways that carry the signal through the cytoplasm and into the nucleus. N-, H-, and K-RAS are the three members of the family. They are GTP binding proteins such that when they are bound to GTP they are activated and when they are bound to GDP they are inactivated. Nucleotide exchange factors, such as SOS mentioned above, catalyze the exchange of GTP for GDP. GTPase activating proteins (GAPs) catalyze the hydrolysis of GTP to GDP to terminate the signal, although RAS proteins possess some intrinsic GTPase activity which allows for self-regulation.

In the EGFR pathway, the protein–protein interactions facilitated by the SH3 domains of Grb2 with the SH3 domains of SOS, bring SOS to the membrane where RAS is located and this results in RAS activation. RAS proteins undergo a series of post-translational modifications that direct their trafficking in the cell. Farnesylation, the addition of the C15 farnesyl isoprenoid lipid to the C-terminal CAAX motif (where C represents cysteine, A represents an aliphatic amino acid and X represents any amino acid), is one modification that is required for localizing RAS to the cell membrane (Figure 4.4). It is interesting to note that it has recently been demonstrated that endogenous RAS is capable of activating downstream signaling pathways from subcellular membrane compartments (ER and Golgi) upon EGF stimulation (Hingorani and Tuveson, 2003). This adds a new dimension to our knowledge since it was previously

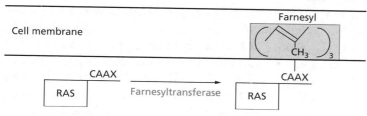

Figure 4.4 Localizing RAS to the cell membrane by farnesylation

thought that localization to the plasma membrane was essential for RAS activity. Furthermore, it has been demonstrated that RAS can transform cells from their subcellular compartments as well as from the membrane. These new observations must be considered in the rationale for designing new cancer therapies.

Raf activation

RAS-GTP binds to and contributes to the activation of the serine/threonine kinase, Raf, one of its main effectors. The recruitment of Raf to the cell membrane is necessary for its activation by tyrosine phosphorylation. Activated Raf phosphorylates mitogen-activated protein kinase-kinase (MAPKK, MEK).

A little lesson about the MAP kinase family: MAP kinase kinase kinases, MAP kinase kinases, and MAP kinases

The nomenclature may seem confusing at first but it is due to the fact that a series of phosphorylation steps is necessary for enzyme activation: a kinase (MAPKKK) phosphorylates another kinase (MAPKK), which itself phosphorylates yet another kinase (MAPK). Raf is a MAPKKK. A unique feature of MAP kinase activation is that it requires both tyrosine and threonine phosphorylation. MAPKK is a dual specificity kinase that phosphorylates both tyrosine and threonine residues.

The MAP kinase cascade

The activated MAPKKs go on to phosphorylate another family of serine/threonine kinases, the mitogen-activated protein kinases (MAPKs) (also known as: extracellular signal regulated kinases; ERKs). The MAPKs are a family of serine/threonine kinases that provide the cytoplasmic link between the activated RAS on the plasma membrane and regulation of gene expression since activated MAPKs can enter the nucleus. The activity of many transcription factors is regulated by phosphorylation and thus MAPKs can affect the activity of transcription factors via phosphorylation.

Although this discussion focuses on MAPK, there are actually three distinct but parallel MAP kinase pathways: MAPK, JNK, and p38. As noted in this section, MAPK is activated by growth factors. JNK and p38 are activated by a various environmental stress signals such as ultraviolet and ionizing radiation. The JNK and p38 pathways usually trigger apoptosis. Thus all three MAPK pathways act as a common mechanism that serves multiple signaling pathways and results in various cellular responses.

Regulation of transcription factors

The AP1 transcription factor is an important target of the MAPK cascade. As a transcription factor, it binds to DNA and regulates the expression of genes involved in growth, differentiation, and death. One mechanism, whereby AP-1 induces cell cycle progression, is by binding to and activating the cyclin D gene, a critical regulator of the cell cycle. AP-1 is not a single protein but rather is made up of the products of two gene families, *jun* and *fos*. These proteins contain basic leucine zipper domains that facilitate their binding as dimers to either the cAMP response element (CRE) or the 12-O-tetradecanoylphorbol-13-acetate (TPA) response element in a target gene. AP-1 activity is induced by two mechanisms. First, direct phosphorylation of members of the Fos family by MAPK affects their DNA binding activity. Secondly, MAPK phosphorylation and subsequent activation of other transcription factors increases the expression of both fos and jun genes. As a result, AP1 activity increases and subsequent transcriptional regulation proceeds.

The Myc family of transcription factors (Myc, Max, Mad, Mxi) can dimerize in different ways and lead to distinct biological effects of growth, differentiation, and death. Several seem to be targets of MAPKs. Myc is a short-lived protein that promotes proliferation by regulating the expression of specific target genes. Gene targets of myc include *N-Ras* and *p53*, but the identification of additional targets is the subject of ongoing research. Myc requires the constitutively expressed family member Max to function. Myc and Max form heterodimers via basic helix–loop–helix leucine zipper domains and bind to a regulatory sequence called the E-box in their target genes. Heterodimer formation and DNA binding are crucial for the oncogenic, mitogenic, and apoptotic effects of Myc. Other heterodimers, such as Max and Mad/Mxi, are inhibitory for Myc function. They can also bind to the E-box in gene promoters but they repress transcription. Thus, the Myc family of transcription factors forms a network of interacting basic helix–loop–helix leucine zipper proteins and the identity of the members within a heterodimer determines the biological effect elicited.

PAUSE AND THINK

The three-letter language of molecular terms sounds peculiar at first but with practice will become familiar and allow you to build your molecular vocabulary.

Self test   Close this book and try to redraw Figure 4.2. Check your answer. Correct your work. Close the book once more and try again.

Good! Now let us backtrack to illustrate how you can build levels of complexity on the foundation you have learned. RAS has been noted to be a "major player" for the integration of a growth factor signal with a number of crucial signaling pathways. In fact, all receptor tyrosine kinases activate RAS. Raf was described above as one effector protein that leads to the activation of the MAPK cascade but there are several other effector proteins of RAS activation. Phosphatidylinositol 3-kinase (PI3K), a lipid kinase, is another effector protein downstream of RAS that can be introduced at this point.

Crystal studies have shown that RAS interacts directly with the catalytic structure of PI3K. Production of the second messenger, PIP3, recruits the serine/threonine kinase PDK-1 to the membrane. Akt, another serine/threonine kinase, is also recruited to the membrane where it is phosphorylated and activated by PDK-1. Activated Akt is translocated into the nucleus where it phosphorylates nuclear substrates, including transcription factors. Activated Akt is also involved in anti-apoptotic and survival roles by phosphorylating distinct target proteins.

Self test Draw a diagram of a growth factor signal transduction pathway including two effector proteins of RAS activation. Check your answer with Figure 4.5. Correct your work.

The effect of cell signaling on cell behavior

We have seen above that growth factor signaling can lead to cell proliferation via signaling to the nucleus. In addition, cell signaling can have effects on cell behavior. This is clearly demonstrated by the intracellular tyrosine kinase, SRC (the gene *src* is discussed below). In addition to having a role in cell proliferation, SRC plays an important role in the regulation of cell adhesion, invasion, and motility upon EGF

PAUSE AND THINK

What is a lipid kinase? It is an enzyme that phosphorylates lipids. PI3K phosphorylates the 3'-OH group of PIP2 (phosphatidylinositol-4,5-bisphosphate) to produce the second messenger, PIP3 (phosphatidylinositol-3,4,5-trisphosphate). Do not be discouraged by these cumbersome names: the difference between the two is only a phosphate group.

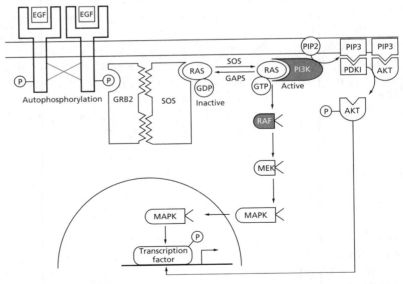

Figure 4.5 The EGF signal transduction pathway showing two effectors of RAS (shown in red)

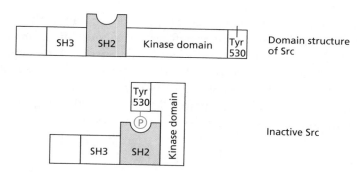

Figure 4.6 The protein domains of SRC and a negative-regulatory intramolecular interaction

receptor activation. SRC is a phosphoprotein which contains SRC homology domains, including a SH2 and SH3 domain. It also contains a negative-regulatory domain near the carboxyl terminus. When Tyr 530 in this negative-regulatory domain is phosphorylated, it binds to the internal SRC SH2 domain and results in an intra-molecular association which represses the kinase domain and keeps SRC in an inactive state (Figure 4.6). One way that SRC can be activated is via receptor tyrosine kinases, such as the EGF receptor. Upon stimulation of the EGF receptor by growth factor, the autophosphorylated receptor can interact with the SH2 domain of SRC, disrupting its negative-regulatory intramolecular conformation. The activated kinase phosphorylates a wide-range of target proteins, including focal adhesion proteins (e.g. focal adhesion kinase, FAK), adaptor proteins, and transcription factors. The regulation of focal adhesion proteins within the dynamic subcellular structures called focal adhesions is particularly important for adhesion and motility. These structures associate with cytoskeletal fibers that ultimately control cell shape and motility. Assembly of focal adhesions facilitates cell adherence while disassembly facilitates motility. Activation of SRC leads to disassembly of focal adhesions and thereby permits increased motility. SRC also regulates cell invasion by inhibiting E-cadherin (see Chapter 8).

Thus, the effects of growth factor cell signaling through tyrosine kinase receptors are indeed multifold, as is illustrated by the many downstream effects that occur upon SRC activation.

4.2 Oncogenes

Cancer arises from mutations in genes that are involved in growth, differentiation, or death. There are two major classifications of mutated genes that contribute to carcinogenesis: oncogenes and tumor suppressor genes. A general description of an oncogene is a mutated gene whose protein product is produced in higher quantities or whose altered product has increased activity and therefore acts in a dominant manner. A mutation in only one allele is sufficient for an effect. Tumor suppressor genes (see Chapter 5) on the other hand, are genes in which the mutation has caused a loss of function and, therefore, most are recessive in nature because both alleles must be mutated. More than 100 oncogenes and at least 15 tumor suppressor genes have been identified.

Studies of retroviruses

Studies of retroviruses have led to great insights into cancer biology and have become the foundation of our knowledge of oncogenes. Several landmark experiments were performed based on the early observation that viruses could cause cancer in animals and the results pointed to the discovery of oncogenes. In 1911, Peyton Rous prepared a cell-free filtrate from a chicken sarcoma and demonstrated that he could induce sarcomas in healthy chickens with this filtrate. The causative agent was identified as the Rous Sarcoma Virus. Many decades later, oncogenic transformation by this virus were found to be due to an "extra" gene contained in its genome, that was not required for viral replication. The first so-called "oncogene" was identified as *v-src* (pronounced v'sark'). The oncogene product was characterized as a 60 kDa intracellular tyrosine kinase.

A leader in the field of oncogenes: Joan Brugge

Joan Brugge has been a personal influence in my own career. Although she was never one of my official mentors, she created an extremely encouraging atmosphere at the State University of New York at Stony Brook while I was carrying out my Ph.D. research. Her style in carrying out research was always natural, enthusiastic, and genuinely inquisitive. She is a model scientist. Joan's own motivation to carry out cancer research stemmed from the loss of a family member to cancer. Joan made great strides early in her career. Working with Ray Erikson at the University of Colorado during her postdoctoral tenure, she carried out pioneering work on the first retroviral oncogene product, the SRC protein.

Joan received her B.A. in biology from Northwestern University and her Ph.D. in virology from Baylor College of Medicine. In addition to the State University of New York at Stony Brook, she was on the faculty at the University of Pennsylvania and is currently Acting Chair of the Department of Cell Biology at Harvard Medical School. In between her academic appointments she was the Scientific Director of ARIAD Pharmaceuticals, Inc. in Cambridge, MA. Joan has recently been awarded a Research Professorship from the American Cancer Society, the Society's most prestigious research award, for her significant contributions to cancer research. She is currently investigating the initiation and progression of breast cancer using three-dimensional cellular models.

In 1976, a startling discovery was made. Bishop and Varmus found that there was a gene with a homologous sequence to *v-src* in uninfected chickens. Moreover, upon further investigation this gene could be found in organisms from fruitflies to humans. Following further examination, a fundamental principle of cancer biology was revealed: almost all known oncogenes are altered forms of normal genes or proto-oncogenes.

■ Exception: some protein products of DNA tumor viruses behave like oncogenes but are not viral versions of cellular oncogenes. They act by blocking the activity of tumor suppressor proteins. See Chapter 5.

The name proto-oncogene is sometimes used in cancer biology to distinguish the normal cellular gene (e.g. *c-src*) from the altered form transduced by retroviruses (*v-src*). The *v-src* sequence lacks the carboxy-terminal negative-regulatory domain present in *c-src* and has point mutations throughout the gene.

A leader in the field of oncogenes: Harold Varmus

Harold Varmus, along with J.M. Bishop, received the Nobel Prize in Physiology or Medicine in 1989 for studies carried out at the University of California, San Francisco, that laid down the foundation for the role of mutations in carcinogenesis. They discovered that some genes of cancer-causing viruses were mutated forms of normal cellular genes. As we see in this chapter, this was the birth of the concept of proto-oncogenes and of the molecular biology of cancer.

Harold Varmus is a native of Long Island, NY (like myself). He obtained a Master's Degree in English from Harvard University and is a graduate of Columbia University's College of Physicians and Surgeons.

Varmus was named by President Clinton to serve as the Director of the National Institute of Health, a position he marked with many advancements during his six years' service. He has acted as an advisor to the federal government and as a consultant for several pharmaceutical companies and academic institutions. He is currently the President and Chief Executive Officer of Memorial Sloan-Kettering Cancer Center in New York City. His current research includes the development of mouse models for human cancer.

At this point a short review of the retroviral life cycle is necessary. The life cycle of retroviruses brand them as intracellular parasites in that they rely on their host cell for energy and to synthesize viral proteins. After injecting their infectious nucleic acid (RNA) into a host cell, the viral RNA is first reverse transcribed into DNA. This provirus DNA is integrated randomly into the host chromosome, where it will be replicated, transcribed, and translated as host DNA. The translation of viral RNA then produces viral proteins for the synthesis of new viral particles. During evolution, the virus can acquire fragments of genes from the host at integration sites and this process may result in the creation of oncogenes. The Rous Sarcoma Virus acquired a truncated form of *c-src*. Alternatively, and depending on the integration site, viral DNA may be translated as a fusion protein in conjunction with cellular DNA resulting in a novel fusion protein, or host genes may fall under the regulation of viral regulatory sequences. The resulting disruptions to host gene expression are other mechanisms of virus-induced oncogenesis. This knowledge aids our general understanding of the mechanisms of carcinogenesis because although viruses are not the major cause of human cancers, the mechanism of oncogenic activation of proto-oncogenes is similar. For example, chromosomal translocations may have the same consequence as the integration of a virus into a host chromosome; a crucial gene may come under the influence of novel regulatory sequences and result in abnormal quantities of the gene product. The new gene configuration may serve as an oncogene.

It is important to become familiar with examples of oncogenes to bolster the lesson learned from viral studies: almost all known oncogenes are altered forms of normal genes.

There are examples of oncogenes for every type of protein involved in a growth factor signal transduction pathway. Several examples are described below.

Growth factors

The first evidence for the role of proto-oncogenes came from analysis of the viral oncogene *v-sis*. Its protein product was cytoplasmic and was found to be a truncated version of a growth factor normally secreted by platelets, called platelet-derived growth factor (PDGF). Thus, the identity of the product of the proto-oncogene or cellular gene *c-sis*, is PDGF. It is a component of a wound response and its normal role is to stimulate epithelial cells around the wound edge to proliferate and repair the damage. The significance of the oncogenic form may be its aberrant location (cytoplasmic rather than secreted) and the subsequent activation of the PDGF signal pathway at inappropriate times (e.g. other than in response to a wound), leading to unregulated growth.

Growth factor receptors

The oncogene, *v-erbB*, was originally identified from (and named after) the avian erythroblastosis leukemia virus. It is a truncated form of the epidermal growth factor receptor whereby the extracellular domain is deleted. Thus, the identity of the product of the proto-oncogene or cellular gene *c-erbB*, is EGFR. This mutated receptor triggers cell division in the absence of EGF. Point mutations that accomplish the same effect of interfering with growth factor binding and inducing constitutive activation have been identified in human cancers. Increasing the amount of normal *c-erbB* product by gene amplification is another mechanism that contributes to carcinogenesis, particularly breast cancer. Gene amplification involves multiple duplications of a DNA sequence due to errors at DNA replication forks.

The proto-oncogene, *ret*, another growth factor tyrosine kinase receptor, heterodimerizes with cell-surface receptors GFR-α1-4 in order to transduce the signal for glial-derived neurotrophic factor (GDNF). It plays an important role in kidney development and neuronal differentiation. Papillary thyroid carcinoma cells often carry somatic chromosomal rearrangements involving the N-terminal parts of numerous genes and the sequences of *ret* that code for the tyrosine kinase domain. The fusion protein products display kinase activity that is independent of GDNF signaling. Germline mutations are associated with three familial tumor syndromes: multiple endocrine neoplasia 2A (MEN2A), MEN2B, and familial medullary thyroid carcinoma.

The mutations that have been identified illustrate different mechanisms for oncogenic activation. Almost all MEN2A patients have mutations in conserved extracellular cysteines. Resulting intermolecular disulfide bonds cause constitutive Ret dimerization and aberrant activation. In MEN2B patients, oncogenic activation is achieved by altering the substrate-binding pocket of the tyrosine kinase domain. A conserved Met is characteristic of the substrate-binding domain of receptor tyrosine kinases whereby Thr is conserved for cytoplasmic tyrosine kinases. The characteristic mutation of MEN2B is a substitution mutation whereby Thr replaces this conserved Met residue (Met918Thr). This results in altered substrate access leading to increased kinase activity and altered substrate specificity that is characteristic of cytoplasmic tyrosine kinases instead of receptor tyrosine kinases. Thus, the signal transduction pathways are highly disrupted.

Oncogenic activation of receptor tyrosine kinases occurs through specific mutations that lead to constitutive ("always on") tyrosine activation or dimerization.

Intracellular signal transducers

The oncogenic activation of *ras* is observed in about 30% of human tumors. The majority of mutations are located in codons 12, 13, and 61. The consequence of each of these mutations is a loss of GTPase activity

PAUSE AND THINK

What type of genes are usually involved in inherited predispositions to cancer? Although tumor suppressor genes are most often involved, here we see an unusual example of an oncogene (*ret*) playing a role in cancer predisposition.

of the RAS protein, normally required to return active RAS-GTP to inactive RAS-GDP. The effect is constitutive activation of RAS protein, even in the absence of mitogens. Some specific mutations in the *ras* gene are characteristic for specific cancers. A point mutation within codon 12 that results in the substitution of valine (GTC) for glycine (GGC) is characteristic of bladder carcinoma while substitution of serine (AGC) is common in lung cancer.

Genes that code for cytoplasmic tyrosine kinases such as SRC, serine/threonine kinases such as RAF and MAPK, and nuclear kinases such as ABL, can also undergo oncogenic activation. As discussed above, intramolecular associations normally regulate c-SRC kinase activity; the SRC SH2 domain binds a C-terminal phosphorylated tyrosine residue (Tyr 530) and results in a conformation that blocks the SRC kinase active site. Repression of SRC kinase activity can be relieved by dephosphorylation of Tyr 530, or by the binding of the SH2 domain to specific activated tyrosine kinase receptors. See Pause and Think. In colon cancer, the protein product of oncogenic *src* is characterized by a truncation at Tyr 530. This aberrant protein is unable to adopt the inactive conformation described above and therefore kinase activity is constitutive ("always on").

Let us look at *abl* as another example. *c-abl* is a gene whose product is a nuclear tyrosine kinase that plays a role in DNA damage-induced apoptosis. It is normally activated by ionizing radiation and particular drugs via the serine/threonine kinase, ATM. Oncogenic activation occurs through the chromosomal translocation t(9;22), whereby *abl* becomes juxtaposed to a breakpoint cluster region, *bcr*. Thus DNA sequences that are normally not next to each other are now fused and, upon transcription, give rise to a fusion protein with novel features (Figure 4.7). Translocated Bcr retains Domain I and II (shown in white) while Abl retains the SH3 and SH2 domains, the kinase domain, the DNA binding

PAUSE AND THINK

How does this interaction described above compare with that of the SH2 domain of Grb2 and a receptor tyrosine kinase? In both cases, the SH2 domain recognizes a phosphotyrosine residue. The SH2 domain of SRC regulates intramolecular interactions but the SH2 domain of Grb2 regulates intermolecular interactions between itself and receptor tyrosine kinases.

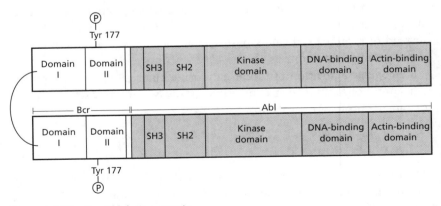

Figure 4.7 The Bcr-Abl fusion protein

domain, and the actin binding domain (shown in red). Bcr-Abl molecules associate with each other forming homo-oligomeric complexes, mediated by the coiled-coil motif in Domain I of Bcr. Oligomerization permits autophosphorylation at Tyr 177 within Domain II, and this triggers activation of the Abl tyrosine kinase. The fusion protein Bcr-Abl is maintained in the cytoplasm. Consequently, a nuclear kinase is activated in the cytoplasm and has access to a range of novel substrates, interfering in the normal signal transduction pathways of the cell.

Transcription factors

It is not surprising that the transcription factor AP-1, which is an important regulator of cell growth, differentiation, and death, is also involved in transformation. Components of AP-1, Jun and Fos, are encoded by proto-oncogenes *c-jun* and *c-fos* and several mechanisms of oncogenic activation of these genes exist. Normally *c-fos* mRNA is short lived so that the response to a mitogen is transient. Truncation of the 3′-end of *v-fos* eliminates a motif involved in mRNA instability (ATTTATTT) and produces an mRNA with a longer half-life. The aberrant expression of *v-fos* mRNA results in an increase in *v-fos* gene product and an inappropriate increase in the transcription of AP1-regulated genes. Oncogenic activation may also involve the deletion of a regulatory promoter sequence, the serum response element, such that transcription of the *fos* gene occurs even in the absence of serum mitogens.

Oncogenic activation of *c-myc* occurs from constitutive and overexpression of the c-Myc protein. Chromosomal translocation of *myc* (chr 8) to a location that falls within the regulation of the strong promoter of immunoglobulin genes (chr14) increases the amount of expression from the *myc* gene. This mechanism of oncogenic activation of *c-myc* is commonly observed in Burkitt's lymphoma. The increase of Myc protein results in an inappropriate increase in the transcription of Myc-regulated genes.

Remember that steroid hormone receptors act as ligand dependent transcription factors. In addition to *v-erbB* discussed above, another oncogene, *v-erbA*, was originally identified from (and named after) the avian erythroblastosis leukemia virus. The identity of the product of the proto-oncogene or cellular gene *c-erbA* is the thyroid hormone (triiodothyronine, T3) receptor. Oncogenic activation is achieved by mutations that prevent thyroid hormone binding and inhibit transcriptional activation. This type of mutation is referred to as a dominant negative mutation because the product of this mutation codes for receptors that can bind to DNA and block access of wild-type receptors, including other steroid hormone receptor family members that may form heterodimers with thyroid receptors at their response elements. Since the product of *v-erbA* can form homodimers (note that the product of the proto-oncogene *erbA* homodimerizes poorly), it is thought that the homodimers mediate the dominant negative effect on the response elements. Most mutations of thyroid hormone receptors identified in human cancers result in dominant negative products suggesting that they may be involved in human cancers, but this is an issue that needs further investigation (Gonzalez-Sancho *et al.*, 2003). See Pause and Think.

Mechanisms of oncogenic activation

As can be seen from the above, several mechanisms can be used to activate proto-oncogenes to become oncogenes (Figure 4.8). Point mutations and deletions in the coding region are a common mechanism and often change the structure and/or function of the proto-oncogene. Both were illustrated for oncogenic activation of the *EGFR* gene. Mutations in the gene promoter region can lead to overexpression of a proto-oncogene. Chromosomal translocations as well as insertional mutagenesis cause the juxtaposition of sequences normally not next to each other and often this configuration can cause altered expression. The translocation involving *c-myc* and immunoglobulin regulatory sequences, which is characteristic of Burkitt's lymphoma, is one example mentioned above. Alternatively, fusion proteins can have novel characteristics. The Philadelphia chromosome relocates the nuclear kinase, c-Abl, to the cytoplasm where it encounters novel substrates. Gene amplification is another mechanism for activation of *erbB* and is observed in breast cancer.

Mechanisms that activate proto-oncogenes to become oncogenes

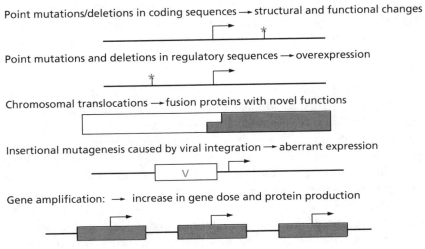

Figure 4.8 Mechanisms that activate proto-oncogenes to become oncogenes

◎ Therapeutic strategies

The knowledge of the molecular details of the EGFR signal transduction pathway and related pathways has led to the launch of many new cancer therapeutics targeting individual components. There has been great success for some and valuable lessons from others. The future is hopeful as we continue to unravel the molecular biology of signal transduction pathways and move forward in the design of additional new therapeutics. The strategies of some new therapeutics aimed at molecular targets within these pathways are described below.

4.3 Kinase inhibitors

Many types of kinases, including transmembrane tyrosine kinases, cytoplasmic kinases and nuclear kinases, are implicated in cancer as we have seen above. Therefore they have become an important target for the design of new cancer therapeutics. In theory, one may predict that designing drugs with great specificity to a subset of kinases would be difficult because the structure of catalytic domains of different kinases are very similar when the kinases are in the active state. However, the synthesis of specific kinase inhibitors has clearly been demonstrated and examples are described below and illustrated in Figure 4.9.

Anti-EGFR drugs

The ErbB2 gene, a member of the EGFR family, is amplified in 30% of breast cancer patients. Overexpression of this gene in cell culture experiments and in transgenic mice (see Box below) results in cell transformation and the induction of breast cancer respectively. This suggests that ErbB2 has a causal role in breast cancer. Several small-molecule kinase inhibitor drugs directed against the tyrosine kinase activity of EGFR family members have been developed, including Iressa (ZD1839), Tarceva, and ABX-EGF. Iressa was approved for the treatment of advanced non-small cell lung cancer (see Pao *et al.*, 2004 and Chapter 10 for further discussion).

Analysis of gene function using transgenic mice

One strategy for investigating the function of a gene is to put it in a place where it is not normally located or remove it from where it normally resides and observe any changes. Transgenic mice contain an additional or altered gene in all of their cells.

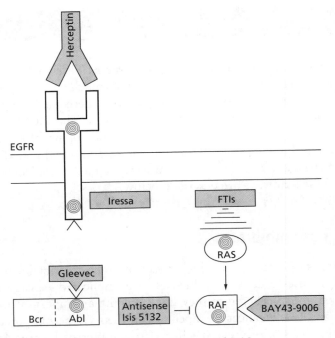

Figure 4.9 Therapeutic strategies that target kinases and RAS

Transgenic mice can be created by the direct injection of DNA into a fertilized egg. More commonly, foreign DNA is introduced into embryonic stem cells in culture prior to their transplantation into an early embryo (blastocyst). A series of breeding stages must follow to generate fully homozygous transgenic mice, since the founder animal is chimaeric (not all cells of the organism contain the altered DNA), rather than transgenic. The cell culture step allows for selection and analysis of DNA integration to take place prior to the creation of a mouse. To create a knockout mouse, a vector is designed so that it will insert into a specific gene location by homologous recombination and disrupt the endogenous gene such that gene function is repressed. Complex transgenic experiments may use tissue-specific or inducible promoters to induce the expression of the foreign gene in a particular location or specific time.

A different approach has also proved successful. Instead of targeting the tyrosine kinase domain of the ErbB2 receptor, the unique extracellular domain was targeted using monoclonal antibodies. Herceptin (Trastuzumab) is a humanized (produced with human recombinant immunoglobulin genes) monoclonal antibody that binds the extracellular domain of ErbB2 with high affinity. Herceptin functions through a combination of mechanisms of action including enhanced receptor degradation, inhibition of angiogenesis and recruitment of immune cells, resulting in antibody-dependent cellular cytotoxicity. Herceptin was approved by the Food and Drug Administration in 1998 for treatment of metastatic breast cancer whose tumors overexpress ErbB2 and thus it is the first genomics-based therapy administered selectively, based on the molecular profile of the tumor. Erbitux (Cetuximab) is a chimaeric antibody directed against EGFR which has also been approved (Switzerland 2003; USA 2004). Genomic characterization of tumor DNA is elucidating subsets of molecularly distinct tumor types within a class of cancer (e.g. breast cancer) that will respond differently to targeted treatments. This is a significant step towards individual-tailored treatment, whereby treatment is matched to the molecular make-up of the patient's tumor and promises to increase the success rate of a given drug.

Strategies against Raf

Since 20% of all tumors have activating mutations in *ras*, targeting downstream effectors could prove valuable as a cancer treatment. Several strategies to target the serine/threonine kinase, Raf, one of the main effectors of RAS, have been developed. One strategy has been to synthesize antisense oligonucleotides that can bind to raf RNA. The RNA hybrids are most likely targeted for degradation or block translation and result in a reduction of Raf protein. The reagent ISIS5132 (Isis Pharmaceuticals) is one such agent that has entered clinical trials. Phase I trials showed the reagent to be non-toxic. Phase II trials were disappointing because a reduction in Raf levels was not observed and no efficacy was observed in patients with non-small cell lung carcinoma. Mutations in *ras* are found in 50% of non-small cell lung carcinomas. Another strategy commonly used to target other kinases has also been applied to Raf. A kinase inhibitor directed towards the ATP-binding site of Raf, called BAY43-9006, has entered clinical trials. In order to investigate whether there was modulation of the defined molecular target, Raf, the phosphorylation of Raf targets was monitored. Data demonstrated a reduction in downstream MAPK phosphorylation in the blood of patients receiving well-tolerated oral treatment. Note that it is important to monitor molecular endpoints (MAPK activity), in addition to clinical endpoints (anti-tumor activity). Results from Phase II and III trials that will evaluate anti-tumor activity are forthcoming.

STI-571 (Gleevec™)

Chronic myelogenous leukemia (CML) accounts for 15–20% of all leukemias. Bone marrow transplantation is the only hope for a cure but is not feasible for a variety of reasons (including donor matching) for the majority of patients. Interferon-α, accompanied by severe side effects, was until recently the standard treatment. The knowledge of the molecular biology of the disease has led to successful specific molecular targeting and the development of a most successful drug. Most CML patients (95%) carry the Philadelphia chromosome, the product of the chromosomal translocation t(9:22)(q34;q11) generating the Bcr-Abl fusion protein. As a result of this translocation, the tyrosine kinase activity of Abl is constitutive and is retained in the cytoplasm rather than the nucleus. As a result of aberrant kinase signaling, there is abnormal proliferation of white blood cells, the hallmark of leukemia. Transformation is dependent on the Bcr-Abl kinase activity and therefore provides the perfect therapeutic target.

STI-571 (Gleevec™), a small-molecule tyrosine kinase inhibitor, has been successful in the treatment of CML, resulting in remission of 96% of early-stage patients. It is a paradigm for targeted cancer therapy, having flown through clinical trials and approval (2001) within three years (discussed further in Chapter 10). The compound was modeled and synthesized after related lead compounds (compounds that shows a desired activity, e.g. kinase inhibition) called phenylaminopyrimidines, identified from high-throughput screens of chemical libraries. The compound was originally optimized for inhibiting PDGF-R tyrosine kinase activity but was later found to inhibit Abl and c-kit as well. STI-571 binds to the ATP binding pocket within the catalytic domain, but the fairly narrow specificity of the compound seems to be due to preferential binding of the drug to the inactive state of the kinase as evidenced by analysis of crystal structures (Schindler *et al.*, 2000). STI-571 recognizes the auto-inhibitory conformation of the activation loop of the protein that regulates the kinase activity. The structure of the inactive state is distinctive between different kinases. The drug has a half-life of approximately 15 h and conveniently allows daily oral administration.

Preclinical data demonstrated inhibition of proliferation in cultured cells and in cells from CML patients with the Philadelphia chromosome as well as tumor regression in mice. This evidence allowed progression to clinical trials. The threshold dose for significant therapeutic efficacy was found to be 300 mg in Phase I trials. Parameters of how well the drug works, efficacy endpoints, were measured by the degree of cytogenetic (chromosomal) and hematologic (blood count) response. A complete cytogenetic response was defined as 0% Philadelphia chromosome positive cells in metaphase (partial, 1–35%; minor, 36–65%, minimal, 66–95%, or no response >95% were additional parameters used). Hematologic response is simply graded by white blood cell counts. Importantly, molecular target inhibition was also analyzed. Quantitation of the levels of phospho-CRKL, a Bcr-Abl substrate found in neutrophils, allowed for the assessment of the inhibition of kinase activity and aided in the determination of effective dosage. As mentioned above, it is important to monitor the modulation of the defined molecular target (Bcr-Abl).

Note that the CML has three disease phases: chronic (lasting 3–5 years), accelerated (lasting from 3–9 months), and blast crisis (lasting 3–6 months). Due to an increase of cell proliferation, the number of white blood cells increases as the disease progresses. The effectiveness of STI-571 decreases with advanced disease phase (53% response in accelerated phase and 30% response in blast crisis). Although only 9% of early-stage patients relapsed, 78% of late-stage patients relapsed. The mechanism for the majority of these cases is due to reactivation of the kinase activity due to mutation or Bcr-Abl amplification. Analysis of clinical samples showed that six out of nine patients had a single amino acid substitution at a contact residue identified in the crystal structure (Gorre *et al.*, 2001). These mechanisms suggest that the initial chromosomal translocation is not only important for initiation but also for maintenance of the cancer phenotype and supports the concept of oncogene addiction: the dependence of a cancer cell on a specific oncogene for its maintenance. STI-571 has also been approved to target c-kit in gastrointestinal stromal tumors and additional studies are investigating its use against PDGF-R in glioblastomas. See Pause and Think.

4.4 RAS-directed therapies

As mentioned above, *ras* is often oncogenically activated during carcinogenesis. The enzyme farnesyltransferase is crucial in the post-translational processing of RAS and its subsequent localization to the plasma membrane and thus posed as a rational target for new cancer drugs. The major strategy was to design compounds called farnesyltransferase inhibitors (FTIs) that would compete with the carboxy-terminal CAAX motif of RAS (see Figure 4.9). Although results were promising in preclinical trials performed on mouse models, those in human clinical trials were not as positive. The discrepancy appears to be due to species differences in enzymatic activities: two members of the RAS family, K-RAS and N-RAS, can be modified by another enzyme, geranylgeranyltransferase (GGT) in the absence of farnesyltransferase so that a type of enzymatic "redundancy" is present in humans but not in mice. However, despite the reporting of "no effect" for many farnesyltransferase inhibitor clinical trials, results from leukemia trials are more encouraging although the mechanism of action is not known. Mutational analysis has suggested that farnesylation of RAS is required for biological functions other than membrane targeting, since it is also important for RAS signaling from subcellular compartments. Therefore, further investigations of the role of farnesyltransferase in carcinogenesis are needed.

> **PAUSE AND THINK**
>
> What is different between the types of molecular targets described above? The examples of molecular targets described above include three different types of kinases: a transmembrane receptor tyrosine kinase, a cytoplasmic serine/threonine tyrosine kinase, and a nuclear tyrosine kinase respectively.

Conclusion

Cancer is a disease characterized by uncontrolled growth. Therefore a clear understanding of growth regulation has helped to reveal the mechanisms of carcinogenesis. This was illustrated by the elucidation of the existence of oncogenes, which include altered versions of normal genes involved in growth. Oncogenes often play a role in growth factor signal transduction. The knowledge of the intricacies of growth factor signal transduction pathways have been and will be essential to the design of successful, low toxicity cancer therapeutics designed against molecular targets.

■ **CHAPTER HIGHLIGHTS—REFRESH YOUR MEMORY**

- Growth factors, growth factor receptors, intracellular signal transducers, and nuclear transcription factors play a role in growth factor signal transduction.

- Many growth factor receptors are tyrosine kinases. Kinases phosphorylate specific amino acid residues in target proteins.

- Phosphorylated proteins can be recognized by specific protein domains (e.g. SH2) and thus can serve as a recruitment platform.

- RAS plays a pivotal role in the EGFR pathway; it links activation of tyrosine kinase receptors with downstream signaling pathways.

- Raf, a serine/threonine kinase activated by RAS initiates a cascade of phosphorylations by the MEK and MAP kinases.

- One ultimate destination of signaling initiated by growth factors is the regulation of transcription factors in the nucleus. Another is effecting cell behavior.

- Retroviruses have been instrumental in the elucidation of oncogenes.

- Most oncogenes are altered versions of normal genes.

- Constitutive kinase activation is a common consequence of oncogenic mutations of tyrosine kinase receptors.

- Aberrant subcellular localization is another consequence of oncogenic activation.

- Many molecular components of growth factor signal transduction pathways have been targets for new cancer therapeutics.

- Different domains of tyrosine kinase receptors have been targeted for the development of new cancer therapies.

- The testing of new therapeutics should include an assay for the modulation of the defined molecular target.

■ **ACTIVITY**

(i)   A new oncogene called ''gre'' has been discovered. It is a tyrosine kinase receptor. Propose a likely mechanism of its oncogenic activation. Describe the components of the signal transduction pathway it may activate, drawing upon your knowledge of other known tyrosine kinase receptors. Suggest a therapeutic strategy for designing a new anticancer drug for this new target.

(ii)   Discuss the importance of protein–protein interactions in growth factor signal transduction pathways.

■ **FURTHER READING**

Bennasroune, A., Gardin, A., Aunis, D., Cremel, G. and Hubert, P. (2004) Tyrosine kinase receptors as attractive targets of cancer therapy. *Crit. Rev. Oncol. Hem.* **50**:23–38.

Blume-Jensen, P. and Hunter, T. (2001) Oncogenic kinase signaling. *Nature* **411**:355–365.

Downward, J. (2003) Targeting ras signalling pathways in cancer therapy. *Nature Rev. Cancer* **3**:11–22.

Druker, B.J. (2002) STI571 (Gleevec) as a paradigm for cancer therapy. *Trends Mol. Med.* **8**:S14–S18.

Sawyers, C.L. (2002) Rational therapeutic intervention in cancer: kinases as drug targets. *Current Opin. Gen. Dev.* **12**:111–115.

Schlessinger, J. (2000) Cell signaling by receptor tyrosine kinases. *Cell* **103**:211–225.

Schlessinger, J. (2002) Ligand-induced, receptor-mediated dimerization and activation of EGF receptor. *Cell* **110**:669–672.

Yeatman, T.J. (2004) A Renaissance for Src. *Nature Reviews Cancer* **4**:470–480.

Zwick, E., Bange, J. and Ullrich, A. (2002) Receptor tyrosine kinases as targets for anticancer drugs. *Trends Mol. Med.* **8**:17–23.

■ **SELECTED SPECIAL TOPICS**

Gonzalez-Sancho, J.M., Garcia, V., Bonilla, F. and Munoz, A. (2003) Thyroid hormone receptors/THR genes in human cancer. *Cancer Letters* **192**:121–132.

Gorre, M.E., Mohammed, M., Ellwood, K., Hsu, N., Paquette, R., Nagesh Rao, P. and Sawyers, C.L. (2001) Clinical resistance to STI-571 cancer therapy caused by BCR-ABL gene mutation or amplification. *Science* **293**:876–880.

Hingorani, S.R. and Tuveson, D.A. (2003) Ras redux: rethinking how and where Ras acts. *Curr. Opin. Gen. Dev.* **13**:6–13.

Pao, W., Miller, V.A. and Kris, M.G. (2004) 'Targeting' the epidermal growth factor receptor tyrosine kinase with gefitinib (Iressa) in non-small cell lung cancer (NSCLC). *Semin. Cancer Biol.* **14**:33–40.

Schindler, T., Bornmann, W., Pellicena, Miller, W.T., Clarkson, B. and Kuriyan, J. (2000) Structural mechanism for STI-571 inhibition of Abelson Tyrosine Kinase. *Science* **289**:1938–1942.

5

Growth inhibition and tumor suppressor genes

Introduction

The human body has mechanisms exerted by tumor suppressor genes that normally "police" the processes that regulate cell numbers and ensure that new cells receive DNA that has been precisely replicated. Recall from Chapter One that the balance between cell proliferation, differentiation, and apoptosis maintains appropriate cell numbers. Many tumor suppressor gene products act as stop signs to uncontrolled growth and therefore may inhibit the cell cycle, promote differentiation, or trigger apoptosis. If both copies of a tumor suppressor gene become inactivated by mutation or epigenetic changes, the inhibitory signal is lost, and the result may be unregulated cell growth, a hallmark of cancer. Other tumor suppressor gene products are involved in DNA repair. If inactivated, DNA repair may be defective and failure to repair DNA may give rise to mutations that lead to cancer. Two alleles of every gene are present in the human genome (except those on sex chromosomes) and, in most cases, loss of tumor suppressor gene function requires inactivation of both copies. This often happens by mutation in one copy and loss of the remaining wild-type allele (Loss of Heterozygosity, LOH).

PAUSE AND THINK

Theoretically, mutation in only one allele allows the other allele to make the tumor suppressor protein and tumor suppression may still occur.

Hereditary syndromes that predispose individuals to cancer can be explained by the inheritance of a germline mutation (passed on from egg/sperm DNA and thus present in all cells of an individual) in one tumor suppressor allele and the acquisition of a somatic mutation or other inactivating alteration in the second allele later in life. This was first proposed by Knudson and is known as Knudson's two-hit hypothesis. It states a strict definition of a tumor suppressor gene: a gene in which a germline mutation predisposes individuals to cancer. Although this hypothesis describes the mechanism by which mutation of most tumor suppressor genes has an effect, exceptions and additional complexities exist and will be mentioned later. A sample of tumor suppressor genes that fit this definition is shown in Table 5.1.

Table 5.1 Tumor suppressor genes. Reprinted from Curr. Opin. Gen. Dev., Vol. 10, Macleod, K., "Tumor suppressor genes", pp. 81–93, Copyright (2000), with permission from Elsevier

Tumor suppressor gene	Human chromosomal location	Gene function	Human tumors associated with sporadic mutation	Associated cancer syndrome	Tumor phenotype of KO mouse mutants (hetero/homozygote)
RB1	13q14	Transcriptional regulator of cell cycle	Retinoblastoma, osteosarcoma	Familial retinoblastoma	MTC, pituitary adenocarcinoma, pheochromocytomas
Wt1	11p13	Transcriptional regulator	Nephroblastoma	Wilms tumor	None
p53	17q11	Transcriptional regulator/growth arrest/apoptosis	Sarcomas, breast/brain tumours	Li–Fraumeni	Lymphomas, sarcomas
NF1	17q11	Ras-GAP activity	Neurofibromas, sarcomas, gliomas	Von Recklinghausen neurofibromatosis	Pheochromocytomas, myeloid leukemia, neurofibromas in DKO chimeras
NF2	22q12	ERM protein/cytoskeletal regulator	Schwannomas, meningiomas	Neurofibromatosis type 2	Sarcomas: metastases on p53 background
VHL	3p25	Regulates proteolysis	Hemangiomas, renal, pheochromocytoma	Von-Hippel Lindau	None
APC	5q21	Binds/regulates β-catenin activity	Colon cancer	Familial adenomatous polyposis	Intestinal polyps in ApcMin
INK4a[a]	9p21	p16Ink4a cdki for cyclinD/cdk4/6; p19ARF binds mdm2, stabilizes p53	Melanoma, pancreatic	Familial melanoma	Lymphomas, sarcomas
PTC	9q22.3	Receptor for sonic hedgehog	Basal cell carcinoma, medulloblastoma	Gorlin syndrome	Medulloblastomas

Gene	Location	Function	Sporadic tumors	Hereditary syndrome	Mouse knockout phenotype
BRCA1	17q21	Transcriptional regulator/DNA repair	Breast/ovarian tumors	Familial breast cancer	None
BRCA2	13q12	Transcriptional regulator/DNA repair	Breast/ovarian tumors	Familial breast cancer	None
DPC4	18q21.1	Transduces TGF-β signals	Pancreatic, colon, hamartomas	Juvenile polyposis	Cooperates with Apc$^{\Delta716}$ in colorect carcinoma
FHIT	3p14.2	Nucleoside hydrolase	Lung, stomach, kidney, cervical carcinoma	Familial clear cell renal carcinoma	Not reported
PTEN	10q23	Dual specificity phosphatase	Glioblastoma, prostate, breast	Cowden syndrome, BZS, LDD	Lymphoma, thyroid, endometrium, prostate
TSC2	16	Cell-cycle regulator	Renal, brain tumors	Tuberous schlerosis	Not reported
NKX3.1	8p21	Homeobox protein	Prostate	Familial prostate carcinoma	Not reported
LKB1	19p13	Serine/threonine kinase	Hamartomas, colorectal, breast	Peutz–Jeghers	Not reported
E-Cadherin	16q22.1	Cell adhesion regulator	Breast, colon, skin, lung carcinoma	Familial gastric cancer	Dominant negative, promotes invasion/metastasis
MSH2	2p22	*mut S* homolog, mismatch repair	Colorectal cancer	HNPCC	Lymphoma, colon/skin carcinoma
MLH1	3p21	*mut L* homolog, mismatch repair	Colorectal cancer	HNPCC	Lymphoma, intestinal adenoma/carcinoma
PMS1	2q31	Mismatch repair	Colorectal cancer	HNPCC	None
PMS2	7p22	Mismatch repair	Colorectal cancer	HNPCC	Lymphoma, sarcoma
MSH6	2p16	Mismatch repair	Colorectal cancer	HNPCC	Lymphoma, intestinal adenomas/carcinomas

a This table does not include the susceptibility genes associated with ataxia-telangiectasia (ATM/ATR), xeroderma pigmentosum (nucleotide excision repair genes) Bloom's syndrome (BLM), Werner's syndrome (WRN) or Fanconi's anemia (FAA, FAC, FAD) although mutation of these genes is associated with cancer predisposition. Nor does it include putative tumor suppressor genes which are subverted by chromosomal translocation, for example, *PML*. Genes such as *MADR2*, TGF-β receptor 2, *IRF-1*, *p73*, *p33ING1*, *PPARγ*, *BUB1* and *BUBR1* have been shown to be mutated in certain human tumors but are not included here because germline mutation of these genes is not yet associated with any hereditary human cancer syndrome. BZS, Bannayan–Zonana syndrome; Ldd, Lhermitte–Duclos syndrome.

Historically, tumor suppressor genes were called "antioncogenes" since some of them seemed to "undo" pathways of oncogene activation. Although the term is no longer used, it can be a helpful tool for illustrating the function of some tumor suppressor genes. The role of aberrant phosphorylation by kinases during carcinogenesis was emphasized in Chapter 4.

It is therefore predictable that some genes that encode phosphatases which antagonize kinase activity, could act as "antioncogenes". Inactivation of these phosphatase genes by mutation removes the inhibitory signal and the kinase activity becomes unregulated. One gene encoding a phosphatase that is frequently mutated in many cancers is PTEN (phosphatase and tensin homolog on chromosome 10). PTEN codes for a phosphatase with dual specificity: it can act as both a protein and lipid phosphatase. Its role as a lipid phosphatase in oncogenesis is best known. PTEN dephosphorylates the membrane lipid PIP3 (phosphatidyl-inositol-3 phosphate) to form PIP2. This antagonizes (shown by the reversed red arrow) the PI3 kinase pathway (Figure 5.1).

Loss of this inhibitory signal in the mutant phenotype results in a constitutively active PI3 kinase pathway and favors oncogenesis. Note that this gene also fits the tumor suppressor definition above since a germline mutation of PTEN causes Cowden syndrome which predisposes patients to cancer.

Note that not all kinases are oncogenic and not all phosphatases are tumor suppressors. Ataxia telangiectasia mutated (ATM) kinase functions in DNA repair as mentioned in Chapter 2, and plays a role in tumor suppression. Many other examples exist.

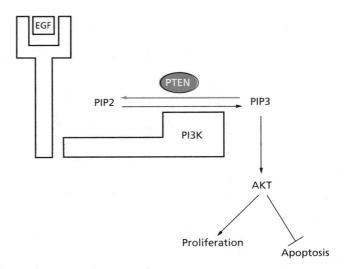

Figure 5.1 PTEN antagonizes the PI3 kinase pathway

An examination of two "star players" in the world of tumor suppressor genes, the retinoblastoma (*Rb*) gene and the *p53* gene, is central to this chapter. The roles of both gene products during carcinogenesis are described below.

5.1 The retinoblastoma gene and protein

Retinoblastoma is a rare childhood cancer with a worldwide incidence of one in 20,000. There are two forms of the disease, a familial (inherited) form and a sporadic form (Figure 5.2). About 40% of all retinoblastoma cases are familial and about 60% are sporadic. In the familial form of the disease, one germline mutation in the *Rb* gene is passed to the child and is present in all cells. A second mutation is acquired in a particular retinoblast that consequently gives rise to a tumor in the retina. One inherited mutated gene results in a sufficiently high probability that a second mutation may occur. It has been suggested that the first mutation generates genomic instability resulting in the high probability of a second mutation and this is sometimes referred to as a mutator phenotype. In sporadic retinoblastoma, both mutations occur somatically in the same retinoblast. Since there are approximately 10^8 retinoblasts in the retina there is a low chance that sporadic retinoblastoma will occur more than once in an individual; hence sporadic cases usually only affect one eye while familial cases are often bilateral. The disease demonstrates Knudson's two-hit hypothesis: two separate mutations—one in each of the

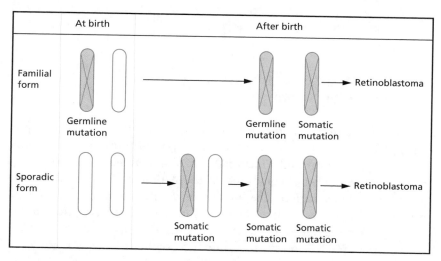

Figure 5.2 The familial and sporadic forms of retinoblastoma: germline vs. somatic mutations

two retinoblastoma alleles—are needed to inactivate the two copies of the Rb allele and prevent expression of the Rb protein. A mutation in one Rb allele is insufficient to knock out functional Rb and so cancer-causing mutations are recessive. Understanding the molecular mechanisms of the gene product underlying this disease elucidates important principles of tumor suppressor proteins.

The retinoblastoma protein (pRB) is the product of the retinoblastoma tumor suppressor gene (*Rb*). Its main role is to regulate the cell cycle by inhibiting the G1 to S phase transition (see Chapter 1). Cell proliferation is dependent on the transcription of a set of target genes to produce proteins that are needed for cell division (e.g. thymidylate synthase; dihydrofolate reductase). pRB is an indirect regulator of transcription for specific gene expression that affects cell proliferation and differentiation. Protein–protein interactions facilitate the function of pRB as a transcriptional regulator; pRB binds to and modulates the activity of a critical transcription factor called E2F and chromatin remodeling enzymes.

PAUSE AND THINK

As a tumor suppressor protein, do you suppose pRB inhibits or activates transcription factors needed for cell proliferation? It inhibits the transcriptional activity of factors needed for cell cycle progression. Therefore, target genes important for cell growth are not expressed. On the other hand, loss of tumor suppressor protein, pRB, results in the loss of inhibition and consequently uncontrolled cell cycle progression and division. Think about the role of pRB in differentiation. Do you suppose it inhibits or activates transcription factors that are responsible for turning on cell-type specific genes? As a tumor suppressor it stimulates the activity of transcription factors, such as Myo D, that activate genes involved in differentiation. Loss of pRB leads to an increase in cell number and to the failure of differentiation.

Structure of the Rb protein

The nuclear retinoblastoma protein, along with two other related proteins p107 and p130, are members of the "pocket proteins" and contain conserved structural and functional domains that bind to various cell proteins. The pocket is composed of the A domain and the B domain joined by a linker region. The binding of the two main cellular effector proteins, histone deacetylase (HDAC) and the E2F transcription factor, to the pocket region (Figure 5.3) is important for its function. HDACs contain the LXCXE motif, an amino acid sequence that is required to bind to the B domain of the pocket of pRB. E2F can bind to the pocket of pRB simultaneously with HDAC because E2F recognizes a different

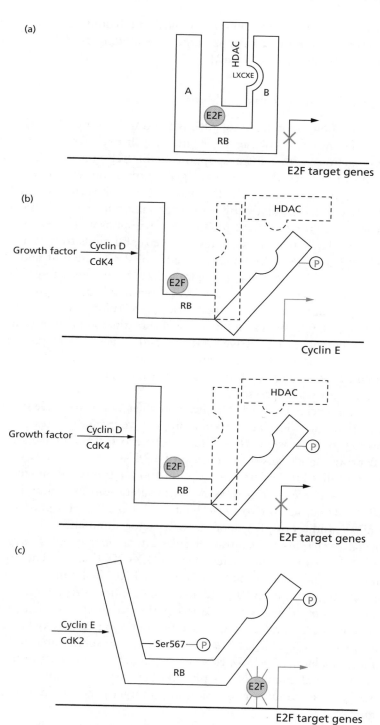

Figure 5.3 Structure and function of the pRB

conserved sequence at the interface of the A and B domains of the pocket. The next question to examine is "how does binding of HDAC and E2F contribute to the function of pRB?"

Molecular mechanisms of the effects of Rb

The major point of control for pRB is the transition from the G1 phase of the cell cycle to S phase (see Figure 1.3). It executes this control by its interactions with the transcription factor E2F and HDACs. (Recall that histone deacetylases regulate gene expression via an epigenetic mechanism; see Chapter 3.) The interactions between pRB and E2F and HDACs are regulated by serine/threonine phosphorylation. In the absence of a growth signal, pRB is in a hypophosphorylated state (i.e. it does not have many phosphates attached) and binds to both E2F and HDAC (Figure 5.3a). By binding to E2F, pRB sequesters it and blocks its transactivation domain, preventing E2F from interacting with the general transcription factors (e.g. TATA binding protein). pRB also inhibits the expression of E2F target genes by recruiting HDACs, enzymes that deacetylate histones and increase chromatin compaction. Thus, the trimeric complex of pRB with HDAC and E2F regulate transcription and consequently cell cycle progression; genes such as *cyclin E, cyclin A*, and *cdk 2*, whose products are required for progression through the cell cycle, are not expressed.

It is the cyclin D and E families and their cyclin dependent kinases (cdks) that phosphorylate pRB in a progressive manner, in response to a growth signal. Phosphorylation leads to conformational changes in the pRB protein and causes the release of E2F and HDAC. HDAC is no longer localized to repress transcription and the transcription factor E2F is free to activate genes necessary for proliferation. Phosphorylation of pRB is carried out in two steps. First, cyclin D/cdk4 phosphorylates C-terminal residues of pRB upon growth factor stimulation. The increase in negative charge causes intramolecular interactions with lysine residues (positively charged amino acids) near the LXCXE domain. The resulting conformational change releases HDAC, a LXCXE bound protein, but not E2F (Figure 5.3b). pRB-mediated transcriptional repression of some genes, and not others, is relieved in the absence of HDAC. The *cyclin E* gene, but not the *cyclin A* gene, is expressed upon the release of HDAC from pRB. The cyclin E/cdk2 complex then phosphorylates additional residues, including Ser 567 close to the linker region. This results in a conformational change of the pRB pocket domain causing the release of E2F and subsequent expression of its target genes, such as cyclin A (Figure 5.3c).

In conclusion, the sequential action of these two cyclin/cdk complexes is important. Phosphorylation of pRB by cyclin D/cdk4 is a prerequisite for cyclin E-cdk2 phosphorylation in that it induces the expression of the *cyclin E* gene and uncovers E-cdk2 phosphorylation sites. Subsequent phosphorylation of pRB by cyclin E-cdk causes the release of E2F. It is speculated that additional cyclin/cdk complexes may also be involved.

Self test   Close this book and try to redraw Figure 5.3. Check your answer. Correct your work. Close the book once more and try again.

5.2 Mutations in the RB pathway and cancer

Retinoblastoma is initiated by the loss of both RB alleles. The types of mutations identified are mostly deletions, frameshift, or nonsense mutations that result in the abrogation of RB function as would be predicted from Knudson's two-hit hypothesis. In addition, missense mutations that lie within the pocket domain have been reported. It is of interest that mutations of Ser 567 (described above) have been found in human tumors because, normally, phosphorylation of this amino acid causes the release of E2F. Mutation of Ser 567 may disrupt the regulation usually observed at this site. Since the RB pathway is central in cell cycle regulation, tumor initiation may be induced via any mutation that blocks Rb function and causes E2F to be available to activate transcription regardless of the presence or absence of a growth signal. Although the RB gene is expressed in all adult tissues, only retinoblastoma and very few other types of cancer are initiated by loss of pRB. Yet this pathway is still inactivated in most human tumors and is targeted by human tumor viruses (see Section 5.5). These observations suggest that we have more to learn about the different roles of pRB and the requirements of different cell types. For example, early studies suggest that pRB may have a role in differentiation as well as cell cycle progression in the developing retina.

5.3 The p53 pathway

◎ The *p53* gene was the first tumor suppressor gene to be identified and, since its discovery, scientists have found that the p53 pathway is altered in most human cancers. Two *p53* homologs, *p73* and *p63*, have also been identified but mutations in cancer cells are rare. p53 is at the heart of the cell's tumor suppressive mechanism and thus has been nicknamed the "Guardian of the Genome". It can be activated by many types of "danger signals", such as cell stress and DNA damage, and can trigger

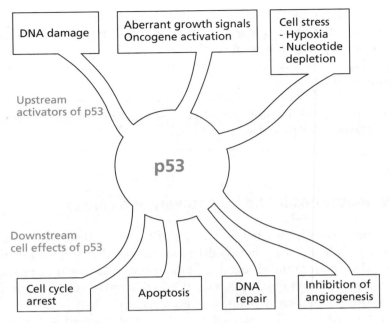

Figure 5.4 Activators and effects of p53

several crucial cellular responses that suppress tumor formation (Figure 5.4). **Upstream** stress activators include radiation-, drug-, or carcinogen-induced DNA damage, oncogenic activation, hypoxia and low ribonucleotide pools. These conditions may nurture tumor initiation. In response to these stress signals p53 can elicit downstream cellular effects including cell cycle arrest, apoptosis, DNA repair, and inhibition of angiogenesis. The ability to cause the cell cycle to pause allows for repair of DNA mutations and prevents their propagation within the genome. Apoptosis is another means of preventing propagation of mutations; cell suicide benefits the organism as a whole if DNA damage cannot be repaired. Mutated cells are better dead. Apoptosis is the critical biological function mediating the tumor suppressor function of p53.

Self test   Close this book and try to redraw Figure 5.4. Check your answer. Correct your work. Close the book once more and try again.

The overall regulation of the p53 pathway possesses an extraordinary complexity that compels us to try to unravel each layer. Let us begin by examining the structure of the p53 protein and its interactions with its inhibitors, and then move onto dissecting how its activity is switched on and how it exerts its effects.

Structure of the p53 protein

The *p53* gene, located on chromosome 17p13, contains 11 exons that encode a 53 kD phosphoprotein. The p53 protein is a transcription factor containing four distinct domains: the N-terminus transactivation domain, the DNA binding domain containing a Zn^{2+} ion, an oligomerization domain, and a C-terminus regulatory domain (Figure 5.5). The p53 protein binds as a tetramer to a DNA response element containing two inverted repeats of the sequence 5′PuPuPuC(A/T)-3′(Pu symbolizes any purine base: A or G) in order to regulate transcription of its target genes. Oligonucleotide array experiments have demonstrated that p53 binds to approximately 300 different gene promoter regions, thus suggesting that p53 has a powerful regulatory role. Several of these specific target genes and the mechanism of how they exert their effect will be discussed later in the chapter.

Regulation of p53 protein by MDM2

Normally, the level of p53 protein in a cell is low. The activity of p53 in a cell is regulated at the level of protein degradation, not at the level of expression of the *p53* gene. The MDM2 protein, an ubiquitin ligase, is its main regulator. Ubiquitin ligases are enzymes that attach a small peptide called ubiquitin to proteins, flagging it for proteolysis (enzymatic protein degradation involving cleavage of peptide bonds) in proteosomes. MDM2 modifies the c-terminal domain of p53, and thus targets it for degradation by proteosomes in the cytoplasm. In addition, MDM2 modifies the activity of p53 since it binds to and inhibits the p53 transactivation domain at the N-terminus and transports the protein into the cytoplasm, away from nuclear DNA. Thus p53's activity as a

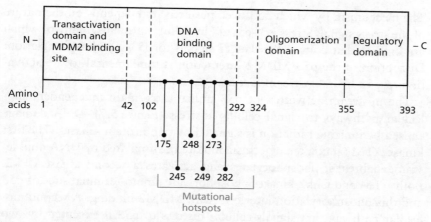

Figure 5.5 Domains of the p53 protein and location of mutational hotspots

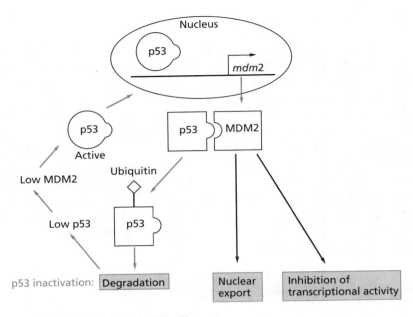

Figure 5.6 Regulation of p53 by MDM2

transcription factor is out of reach. The binding of MDM2 to p53 is part of an autoregulatory feedback loop (Figure 5.6, shown by red arrows) since the MDM2 gene is a transcriptional target of p53. Therefore, p53 stimulates the production of its negative regulator MDM2 that causes the degradation of p53. Low amounts of p53 will reduce the amount of MDM2 protein and this will result in an increase of p53 activity thus completing the loop.

Upstream: molecular pathways of p53 activation

The mechanism by which p53 becomes activated depends on the nature of the stress signal. Stress is "sensed" by cellular proteins, many of which are kinases that convey the danger signals to p53 via phosphorylation. Disruption of the p53-MDM2 interaction is fundamental to the activation of p53 by its upstream factors.

The upstream activators of p53 utilize three main independent molecular pathways to signal cellular distress (Figure 5.7). DNA damage caused by ionizing radiation is signaled by two protein kinases. The first kinase ATM (ataxia telangiectasia mutated), stimulated by DNA double-stranded breaks, phosphorylates and activates a second kinase Chk2. Both ATM and Chk2 kinases phosphorylate amino-terminal sites of p53 and this phosphorylation interferes with MDM2 binding. A second molecular pathway that signals cellular distress to p53 is executed by two different kinases, ATR and casein kinase II. These also phosphorylate

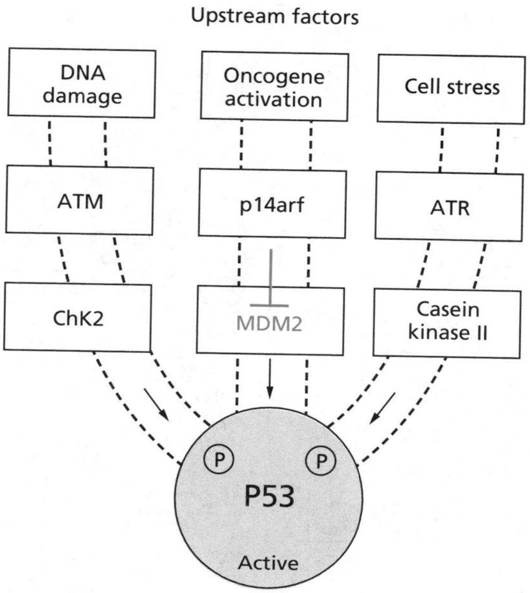

Upstream factors

Figure 5.7 Upstream activators of p53

p53 and disrupt its interaction with MDM2. Lastly, activated oncogenes, such as Ras, induce the activity of the protein p14arf, another modulator of the p53-MDM2 complex. P14arf is one of two translational products of the *INK4a/CDKN2A* gene (p16, a cyclin kinase inhibitor, is the other product). P14arf does not bind to the interface of p53-MDM2, but functions by sequestering MDM2 to the nucleolus of the cell. All three pathways prevent degradation of p53 by MDM2.

Downstream: molecular mechanisms of p53 cellular effects

The main mechanism by which p53 exerts its tumor suppressing effects is by inducing the expression of specific target genes. Let us examine how the resulting network of proteins triggers these responses (Figure 5.8).

Inhibition of the cell cycle

One of the central functions of p53 is to cause cell cycle arrest in response to DNA damage so that there is an opportunity to repair the damage prior to the next round of replication, thus damaged DNA will be prevented from being replicated and passed on to daughter cells and maintenance of the genome will be facilitated. The molecular mechanism responsible for this cellular response involves the transcriptional induction of the *p21* gene. Its product, the p21 protein, inhibits several cyclin-cdk complexes and causes a pause in the G1 to S (and G2 to M) transition of the cell cycle. See Pause and Think.

PAUSE AND THINK

Why would an inhibitor of cyclin-cdk complexes cause a pause in the G1-S transition? Recall the role of cyclin-cdk complexes in the cell cycle; importantly they act as kinases. As kinases they phosphorylate. What do they phosphorylate? pRB. Failure of the cdk complex to phosphorylate pRB prevents the release of the transcription factor E2F and blocks the transition into S phase.

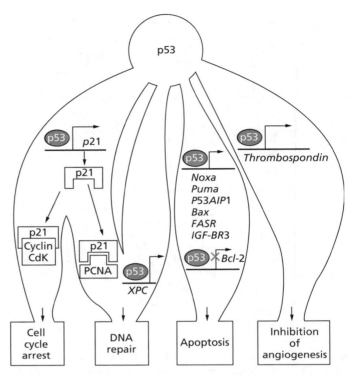

Figure 5.8 Downstream effects of p53

In addition, p21 also binds PCNA (proliferating cell nuclear antigen), a protein that has a role in DNA synthesis and DNA repair. The interaction with p21 is such that it inhibits PCNA's role in DNA replication but not in DNA repair. Therefore, p21 is an important part of the molecular mechanism that facilitates the ability of p53 to bring about a pause in the cell cycle and at the same time allow DNA repair.

Apoptosis

The expression of several mediators of apoptosis is transcriptionally regulated directly by p53 (Table 5.2). The targets include genes that code for proteins involved in two apoptotic pathways that respond to external and internal signals respectively. (Apoptosis will be described in Chapter 6.) In general, genes encoding proteins that promote apoptosis, pro-apoptotic proteins, are induced while genes encoding proteins that antagonize apoptosis, anti-apoptotic proteins, are repressed. The mitochondrial pro-apoptotic proteins NOXA, PUMA, and p53AIP1, that cause the release of cytochrome c and activate the apoptosome, are induced. Also, p53 tips the balance regulated by the Bcl-2 protein family towards apoptosis by inducing gene expression of the pro-apoptotic protein Bax and repressing the expression of anti-apoptotic protein Bcl-2.

Table 5.2 p53-inducible apoptotic target genes

Gene	Location of gene product
Bax	Intrinsic pathway
NOXA	Intrinsic pathway
PUMA	Intrinsic pathway
P53AIP1	Intrinsic pathway
FAS	Extrinsic pathway
IGF-BP3	Extrinsic pathway
DR5	Extrinsic pathway
PIDD	Extrinsic pathway
PERP	Endoplasmic reticulum

Fas receptor (FASR) is a transmembrane receptor that receives extracellular stimuli to stimulate apoptosis. Expression of the *Fas receptor* gene is induced by p53. Apoptosis is also triggered when survival signaling is blocked by the p53 induction of IGF-BP3 (insulin-like growth factor-binding protein 3). IGF-BP3 blocks the signaling of IGF-1 to its receptor. Activation of these different pathways in concert is required for a full apoptotic response. Transcription-independent mechanisms for the induction of apoptosis by p53 also exist and will be discussed in Chapter 6.

DNA repair and angiogenesis

Both DNA repair and angiogenesis are covered in depth elsewhere in this volume (Chapters 2 and 8 respectively). In general, a role for the transcriptional regulation of important genes in these processes by p53 has been established. For example, the gene *XPC* that is involved in nucleotide excision repair is regulated by p53 through a p53 response element in its promoter. Thrombospondin, an inhibitor of angiogenesis, is also transcriptionally regulated by p53. This further supports the role of p53 as a transcriptional regulator in different biological responses.

Decision making

As the guardian of the genome, p53 prevents damaged DNA from being passed on to daughter cells either by inhibiting the cell cycle or by inducing apoptosis. Cell cycle inhibition and apoptosis are two independent effects of p53. The molecular factors that determine the biological outcome of whether inhibition of the cell cycle or apoptosis takes place are just being elucidated. One model that has been put forth is that different combinations of transcription factors that act as dimers influence the biological response. Oncogene activation (e.g. Myc) is an upstream inducer of p53 that triggers apoptosis. The mechanism of this stress signal acts via the cyclin/cdk inhibitor p21, the main effector of cell cycle inhibition but also an inhibitor of cell death. The regulation of the *p21* gene is a pivotal point in the p53 decision-making process. Both p53 and a transcription factor called Miz-1 are required for *p21* gene expression. Now enter the oncogene, Myc, which competes with p53 for binding with Miz-1. Myc interacts with Miz-1 and inhibits the transcription of *p21*. Through this mechanism of preventing expression of *p21*, Myc not only overrides the p53-regulated block to cell cycle progression, but also blocks the p21-mediated inhibition of apoptosis (Figure 5.9). p53 is not altered and is free to induce the expression of pro-apoptotic targets. Additional events are also required for full activation of apoptosis since p53 phosphorylation and apoptotic cofactors are required for the induction of some apoptotic genes. Revealing the mechanisms behind other modes of upstream stress inducers of p53, such as oxidative stress, requires further studies.

The apoptosis stimulating proteins of p53 (ASPP) family also plays a role in p53 decision making (Slee and Lu, 2003). These proteins bind to the p53 DNA binding domain and have been shown specifically to enhance the ability of p53 to activate genes involved in apoptosis and not cell cycle arrest. The selection of apoptotic genes vs. growth arrest genes could potentially be accomplished by specific promoter sequences that

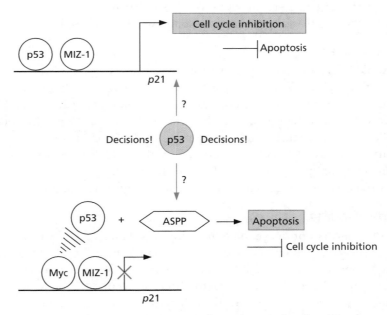

Figure 5.9 Molecular factors in deciding the effects of p53: cell cycle inhibition or apoptosis?

serve to distinguish the functionally distinct classes of genes. The regulation of ASPP itself requires further study. Mutations in the ASPP binding site of the *p53* gene and epigenetic silencing of the *ASPP* gene have been identified in tumor cells. These tumor cells may have been initiated because they escaped from the apoptotic program normally augmented by ASPP. Other co-activators may also enhance the selectivity of p53 to activate apoptotic genes.

5.4 Mutations in the p53 pathway and cancer

Due to the central role of p53 as a tumor suppressor and guardian of the genome, cell transformation is less likely to occur in cells that maintain a functional p53 pathway. p53 mutant cells are characterized by genomic instability since mutations are more likely to be maintained in dividing cells, providing an environment that is permissive for tumor initiation. The high frequency of p53 pathway mutations found in tumor cells is most likely to be the result of selective pressure favoring mutant cells that escape tumor suppression. Over 75% of all *p53* mutations are missense mutations and result in single amino acid substitutions. Many mutant p53 molecules are more stable than wild-type p53 protein and can accumulate in cells. More than 90% of the missense mutations are located in the DNA binding domain (amino acids 102–292) and more than 30% of these affect only six codons and are therefore referred to as "hotspots" (Figure 5.5).

In addition to mutation of the *p53* gene, there are other ways to interfere with the p53 pathway. Defects in pathways that lead to the activation of p53 in response to stress, such as mutations in the *chk2* gene, have been identified in cancer cells that do not contain mutations in the *p53* gene. In these cases, the stress signal would not be transmitted to p53 via phosphorylation as in normal cells (see Figure 5.7). Also, overexpression of the MDM2 protein has been demonstrated to alter the regulation of p53 leading to a "p53-inactivated" phenotype. Inactivation of downstream effectors, such as Bax, also perturb the pathway. Some of these other p53 pathway disruptions may not lead to effects as severe as *p53* mutation, since p53 protein is a central node for receiving and eliciting many stress signals and biological responses respectively, but may mimic a partial aspect of p53 inactivation and be permissive for tumor formation.

Li-Fraumeni syndrome

Li-Fraumeni syndrome is predominantly characterized by a germline mutation of the *p53* gene and leads to a predisposition to a wide range of cancers. It is an autosomal dominant disease so an affected individual has a 50% chance of passing the mutation to each offspring. Patients have a 25-fold increased risk of developing cancer before they are 50 years old compared with the general population. The young age at which individuals develop cancer and the frequent occurrence of multiple primary tumors in individuals are characteristic features of the syndrome. The types of cancer seen within families that carry the mutation include sarcomas, breast cancer, leukemia and brain tumors. Cancer develops at an earlier age over several generations. The protein product of the mutated *p53* gene does not function to protect the genome and an accumulation of mutations goes unchallenged and eventually leads to tumor formation.

More complex than Knudson's two-hit hypothesis

The mechanism of tumor suppressor genes may be more complex than Knudson's two-hit hypothesis suggests. This is particularly clear for *p53*. Loss of heterozygosity is not commonly observed in tumors from cells with only one p53 allele suggesting reduced amounts (haploinsufficiency) of p53 can cause transformation. Mutations may result in varied amounts of tumor suppressor gene expression and therefore tumor suppressor "dose" may play a role in the cancer outcome. Recent experiments which generated p53 hypomorphs (animals that exhibit reduced levels of p53 expression) using RNA interference (see Section 1.5) support this mechanism (Hemann *et al.*, 2003).

Unlike most other tumor suppressor genes, some *p53* mutations do not lead to loss of function. Some missense mutations form an altered protein that acts with the product of a normal p53 allele via dimerization to inactivate its function. This type of effect is referred to as dominant negative, whereby the mutated gene product dominates to inactivate the wild-type gene product. Under this situation, the autoregulatory loop is affected because p53 fails to induce its inhibitor, MDM2, and as a result p53 mutant protein accumulates. Other mutations can lead to a newly acquired 'gain of function' phenotype that can accumulate in and transform cells. In this instance, mutant *p53* (not wild-type *p53*) can be considered an oncogene having an active role in carcinogenesis.

5.5 Interaction of DNA viral protein products with Rb and p53

Viruses are cellular parasites that hijack host cell proteins to maintain their life cycle. Coercing the host cell into S phase is essential for viral propagation. Several DNA viruses have been shown to be carcinogenic

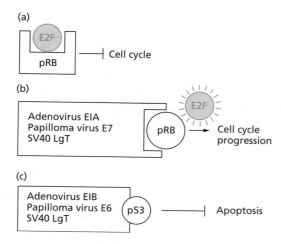

Figure 5.10 Viral protein products interact with p53 and pRB

to humans. Most notably are papovaviruses, adenoviruses, herpes viruses, and hepatitis B. Interestingly, several of the DNA viruses share a common oncogenic mechanism that involves the interaction of viral proteins with the two important regulators of tumor suppression, Rb and p53 (Figure 5.10). The viral proteins: adenovirus E1A, papilloma virus E7, and SV40 Large T antigen inactivate Rb (Figure 5.10b) and SV40 Large T antigen, adenovirus E1B, and papilloma virus E6 inactivate p53 (Figure 5.10c).

The ability of both E6 and E7 to degrade p53 and pRB respectively, using the ubiquitin-proteasome system, correlates with their oncogenic potential. The biochemical events involved in p53 degradation by E6 is as follows: E6 binds to a ubiquitin-protein ligase (E6-AP) and forms a dimer that subsequently binds to p53. p53 is then ubiquitinated and tagged for recognition by the proteosome for degradation. A similar mechanism is likely for E7 targeted degradation of pRB and E1B targeted degradation of p53.

A leader in the field of apoptosis: David Lane

David Lane's contributions to cancer research were acknowledged when he received a knighthood from the Queen of England in January 2000. He discovered the p53 protein-SV40 T antigen complex that helped characterize the mechanisms of viral transformation and tumor suppressor gene function. Sir David was the second most highly cited Medical Scientist in the UK in the last decade. He was instrumental in publicizing the importance of p53 as the "Guardian of the Genome".

David carried out his undergraduate and postgraduate degrees at University College London where he studied auto-immunity. He carried out two postdoctoral tenures, →

> → one at the Imperial Cancer Research Fund (ICRF) in London and the other at the Cold Spring Harbor Laboratory in New York. David returned to the UK to carry out research first at Imperial College, London, and later at the ICRF laboratories. He is currently a Professor and Director of the Cancer Research UK Transformation Research Group at the University of Dundee. David is also the Founder and Chief Scientific Officer of a biotechnology company called Cyclacel, which is developing new cancer therapeutics based on the biology of the tumor suppressor protein p53.

Human papilloma virus (HPV) causes cervical cancer

The *p53* gene is rarely mutated in cervical cancers and suggests that the causative agent, HPV, may be functionally equivalent to *p53* mutation. A polymorphism in the *p53* gene at amino acid 72 in humans leads to differences in the risk of cervical cancer following exposure to HPV. Patients with two alleles coding for Arg at this position have a seven times higher risk of cervical cancer than those with alleles coding for Pro at this site. The Arg containing p53 protein is more susceptible to degradation by HPV E6 (probably due to altered protein conformation). As a result of these cells having decreased p53 activity, they are likely to have an increased mutation rate and an increased potential to form tumors.

 Therapeutic strategies

5.6 Cyclin dependent kinase inhibitors

The Rb pathway is commonly deregulated in cancer. As we have seen above (Section 5.1), phosphorylation and subsequent inactivation of Rb by cyclin dependent kinases is a key step in its regulation. These serine/threonine kinases are overexpressed and/or amplified in some cancers, making them possible molecular targets. A semi-synthetic flavonoid called flavopiridol acts as a competitive inhibitor of all cdks tested by targeting their ATP binding site (Senderowicz, 2001). Flavopiridol induces cell cycle arrest at G1 and G2/M phase. It also affects cdk family members that have a role in transcriptional control and inhibits gene expression of cyclin D1 and D3. Flavopiridol was the first cdk inhibitor to be tested in clinical trials. Anti-tumor activity was demonstrated in some patients but further studies are ongoing. In light of the increasing complex roles of the cdk family (such as in transcriptional regulation and neuronal function), potential side effects of the drug will require careful evaluation. New semi-selective cdk inhibitors (e.g. paullones, oxindoles) are being developed.

5.7 Targeting of the p53 pathway

The role of p53 as a "star player" in suppressing tumorigenesis and the high occurrence of mutations in the *p53* gene found in tumors draws attention to the p53 pathway as a promising cancer therapeutic target. As a result, many different strategies that target the p53 pathway have been developed. Several are described below. The variety of p53 pathway aberrations, including both *p53* gene mutations and defective regulation, suggests that the future success of these therapies will be dependent on knowing the *p53* genotype of tumors in patients prior to treatment. We need to know if there are mutations in the *p53* gene itself or in its regulators and, if so, the type of mutations present. You have to know what is wrong before you can fix it. Therapeutics may strive to correct a *p53* mutation or potentiate normal p53 protein function in cases where other alterations in the p53 pathway affect its function. Several different strategies are described below and illustrated in Figure 5.11 (therapeutics are shaded red).

Strategies that aim to correct a p53 mutation

Gene therapy is one of the most obvious approaches to correct for a *p53* mutation. In fact, many different vectors have been examined in preclinical and clinical settings. Frequently used adenoviral vectors can carry large amounts of DNA and can infect a range of cell types via CAR (coxsackievirus and adenovirus receptors) with high efficiencies. Adenoviruses that have been modified to be replication-defective have been popular vectors. As mentioned above, wild-type adenovirus can replicate in cells by inactivating p53 and pRB (Figure 5.12a). The Onyx 015 virus, a

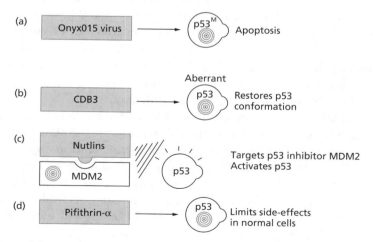

Figure 5.11 Therapeutic strategies that target the p53 pathway

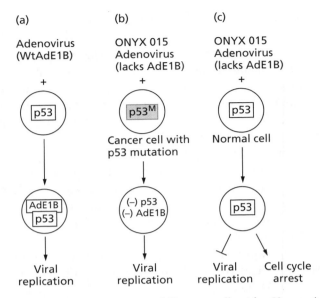

Figure 5.12 The Onyx 015 adenovirus can kill cancer cells with *p53* mutations

replication-selective adenovirus, was designed selectively to kill cancer cells that contain p53 mutations (Figure 5.11a). It takes advantage of the fact that interference with the pRB and p53 pathways is exploited both by viruses and cancer cells (Ries and Korn, 2002). The Onyx 015 virus contains a deletion of the E1B gene and thus it can only replicate within and subsequently kill, cells that have an inactive p53 pathway (Figure 5.12b). The E1A product of the Onyx 015 virus binds to and inactivates pRB and, as a result, induces the G1-S phase transition of the cell cycle allowing for viral replication and cell destruction. The Onyx 015 adenovirus (lacking E1B) triggers a p53 response (growth arrest or apoptosis) in normal cells, resulting in the interference of viral replication (Figure 5.12c). Such selective replication produces a treatment with minimum side effects. Phase I and II clinical trials have given encouraging results. Intratumoral injection proved safe and specific anti-tumor effects were demonstrated. Phase III trials are on-going.

Small molecules that can restore wild-type function to products of *p53* gene mutations have been investigated (Bykov *et al.*, 2003). Many missense mutations of *p53* result in aberrant protein conformation and subsequently interfere with the DNA binding function of the p53 protein. As a result, the p53 inhibitor, MDM2, and target genes essential for apoptosis are not induced. Note that tumors that carry such mutations are more likely to be resistant to conventional chemo- and radiotherapies since these drugs work via the induction apoptosis. Reactivation of p53 mutants aims to eliminate tumor cells via the induction of apoptosis.

A short (nine residues) peptide, CDB3, has been shown to stabilize the structure of several p53 mutants and to restore their transcriptional function (Figure 5.11b). CDB3 is thought to act as a chaperone protein that aids in the refolding of the mutant p53 protein. Similarly, another peptide, peptide 46, stimulates apoptosis when introduced into p53 mutant tumor cells.

Strategies that aim to activate endogenous p53

In many tumors, wild-type *p53* is expressed but the regulation of the p53 protein is defective and results in an altered p53 pathway. Note cancers caused by viruses do not usually harbor a p53 mutation; instead viral proteins act to inactivate p53 function. An approach on the horizon to activate wild-type p53 protein is the development of inhibitors of the p53-MDM2 interaction (Chene, 2003).

Detailed structural information of the p53-MDM2 complex has been obtained by nuclear magnetic resonance and X-ray crystallography data and has revealed that MDM2 has a well-defined binding site for p53. The opposite is not true. Thus, inhibitors are best designed to mimic amino acids of p53. In addition, the binding interface was found to be relatively small which suggested that it was possible to design small inhibitors that could be taken orally. Following this strategy, drugs called nutlins (Figure 5.11c) have been identified and show promising results in preclinical tests (Vassilev *et al.*, 2004). These results include triggering p53 activation and its biological responses in cancer cells containing wild-type p53. In addition, inhibition of tumor growth by 90% was demonstrated in animal models. Nutlins support the idea that protein–protein interactions are good targets for cancer therapeutics.

> **PAUSE AND THINK**
>
> What type of information is required to begin to design such inhibitors?

Strategies that aim to suppress endogenous p53

It is clear that p53 is important for the disposal of tumor cells.

The success of chemo- and radiotherapy is often limited by side effects in normal tissues. Many of the side effects of chemo- and radiotherapy are in part mediated by p53. There is normally high expression of p53 in the tissues that are sensitive to these conventional therapies, such as the hematopoietic organs and intestinal epithelia, and the DNA damage caused by these agents induce p53 to elicit apoptosis, the mechanism behind the side effects. Therefore, suppression of p53 in normal tissue may help alleviate side effects of conventional therapies only in patients with tumors that have lost p53 function. A chemical screen has identified pifithrin-α (p fifty three inhibitor) as a potential agent to test this approach (Figure 5.11d). Prevention of hair loss and an increase in tolerated dose in irradiated mice show promise in preclinical tests.

> **PAUSE AND THINK**
>
> When and where would suppression of p53 be clinically beneficial?

■ CHAPTER HIGHLIGHTS—REFRESH YOUR MEMORY

• Tumor suppressor genes act as STOP signals for uncontrolled growth.

• *PTEN* is a tumor suppressor gene that codes for a phosphatase regulating the activity of a potential oncogenic kinase.

• Knudsons's two-hit hypothesis is demonstrated by the role of the Rb tumor suppressor gene in retinoblastoma: mutations in both alleles are necessary for tumor initiation.

• The retinoblastoma tumor suppressor regulates transcription of cell cycle genes by protein–protein interactions with the E2F transcription factor and HDACs.

• The activity of pRB is regulated by phosphorylation via different cyclin/cdks.

• Hypophosphorylated pRB inactivates E2F and recruits HDACs.

• Phosphorylated pRB releases E2F and HDACs, which facilitates transcription and cell cycle progression into S phase.

• The tumor suppressor p53 has been nicknamed the Guardian of the Genome because of its central role in maintaining the integrity of the cell's DNA.

• The p53 protein is a transcription factor that regulates genes involved in inhibition of the cell cycle, DNA repair, apoptosis and angiogenesis.

• More than 90% of p53 missense mutations are located in the DNA binding domain.

• MDM2 is a main regulator of p53 protein activity.

• The protein product of the p21gene, a cdk inhibitor, is key for eliciting the p53 response of cell cycle inhibition.

• Several gene products, including Bax and IGF-BP3, are important for eliciting the apoptotic response of p53.

• The biological response exerted by p53, either inhibition of the cell cycle or apoptosis, is mediated by the regulation of p21 and by the ASPP family.

• Li-Fraumeni syndrome is a disease that is characterized by an inherited mutation of the *p*53 gene. Patients have a predisposition to a variety of cancers.

• Viral proteins from adenovirus, papilloma virus, and SV40 virus inactivate p53 and pRB as a common oncogenic mechanism; several utilize the ubiquitin-proteosome system.

• Both the pRB pathway and the p53 pathway provide molecular targets for the design of new cancer therapeutics.

■ ACTIVITY

Choose a genetic syndrome that leads to a predisposition to cancer (excluding Li-Fraumeni syndrome). Describe in detail the molecular mechanisms involved.

■ FURTHER READING

Adams, P.D. (2001) Regulation of the retinoblastoma tumor suppressor protein by cyclin/cdks. *Biochim. Biophys. Acta* **1471**:M123–M133.

DiCiommo, D., Gallie, B.L. and Bremner, R. (2000) Retinoblastoma: the disease, gene and protein provide critical leads to understand cancer. *Semin. Cancer Biol.* **10**:255–269.

Guimaraes, D.P. and Hainaut, P. (2002) TP53: a key gene in human cancer. *Biochimie* **84**:83–93.

Lane, D.P. and Lain, S. (2002) Therapeutic exploitation of the p53 pathway. *Trends Mol. Med.* **8**:S38–S42.

Levitt, N.C. and Hickson, I.D. (2002) Caretaker tumour suppressor genes that defend genome integrity. *Trends Mol. Med.* **8**:179–186.

Macleod, K. (2000) Tumor suppressor genes. *Curr. Opin. Genetics Dev.* **10**:81–93.

Ryan, K.M., Phillips, A.C. and Vousden, K.H. (2001) Regulation and function of the p53 tumour suppressor protein. *Curr. Opin. Cell Biol.* **13**:332–337.

Sherr, C.J. (2004) Principles of tumor suppression. *Cell* **116**:235–246.

Sulis, M.L. and Parsons, R. (2003) PTEN: from pathology to biology. *Trends Cell Biol.* **13**:478–483.

Swanton, C. (2004). Cell-cycle targeted therapies. *The Lancet Oncol.* **5**:27–36.

Volgelstein, B., Lane, D. and Levine, A. (2000) Surfing the p53 network. *Nature* **408**:307–310.

■ WEB SITES

International Agency for Research on Cancer; Database of p53 mutations
http://www.p53.iarc.fr/index.html

■ SELECTED SPECIAL TOPICS

Bykov, V.J.N., Selivanova, G. and Wiman, K.G. (2003) Small molecules that reactivate mutant p53. *Eur. J. Cancer* **39**:1828–1834.

Chene, P. (2003) Inhibiting the p53-MDM2 interaction: an important target for cancer therapy. *Nature Rev./Cancer* **3**:102–109.

Hemann, M.T., Fridman, J.S., Zilfou, J.T., Hernando, E., Paddison, P.J., Cordon-Cardo, C., Hannon, G.J. and Lowe, S.W. (2003) An epi-allelic series of p53 hypomorphs created by stable RNAi produces distinct tumor phenotypes *in vivo*. *Nature Genet.* **33**:396–400.

Ries, S. and Korn, W.M. (2002) ONYX-015: mechanisms of action and clinical potential of a replication-selective adenovirus. *Br. J. Cancer* **86**:5–11.

Senderowicz, A.M. (2001) Development of cyclin-dependent kinase modulators as novel therapeutic approaches for hematological malignancies. *Leukemia* **15**:1–9.

Slee, E.A. and Lu, X. (2003) The ASPP family: deciding between life and death after DNA damage. *Toxicol. Letters* **139**:81–87.

Vassilev, L.T., Vu, B.T., Graves, B., Carvajal, D., Podlaski, F., Filipovic, Z., Kong, N., Kammlott, U., Lukacs, C., Klein, C., Fotouhi, N. and Liu, E.A. (2004) *In vivo* activation of the p53 pathway by small-molecule antagonists of MDM2. *Science* **303**:844–848.

6

Apoptosis

Introduction

As described in Chapter 1, the balance between cell growth, differentiation, and apoptosis affects the net numbers of cells in the body and aberrant regulation of these processes can give rise to tumors. Not only do defects in apoptosis contribute to carcinogenesis but they also influence the effectiveness of conventional therapies that mainly exert their effect by inducing apoptosis. In this chapter we will examine the molecular mechanisms of apoptosis and specific mutations that affect the apoptotic pathway and play a role in carcinogenesis. We will also investigate how mutations in the apoptotic pathway can lead to resistance to chemotherapeutic drugs. Lastly, strategies for the design of new cancer therapeutics that target apoptosis will be presented. Let us begin with a description of apoptosis.

Apoptosis is a highly regulated process of cell death which controls cell numbers and gets rid of damaged cells and therefore plays an important role in tumor suppression. Elimination of cells that have damaged DNA helps protect the entire organism from cancer. The apoptotic process is organized and neat, and sharply contrasts the "sloppy" process of necrosis, whereby cells spill out their contents into the surrounding tissue and cause inflammation. Apoptosis is characterized by cell shrinkage and precise chromatin fragmentation that contributes to the neat disposal of the cell. It is an active process requiring the expression of a genetic program that every cell is capable of executing.

Similar to the central role that kinases have in growth factor signaling pathways, particular proteases, called caspases, play a central role in apoptosis. Proteolysis helps to break down cellular components for the neat disposal that is characteristic of apoptosis. The dying cell is swept clean during phagocytosis by macrophages and neighboring cells that recognize molecular flags (e.g. phosphatidylserine) exhibited by the apoptosing cell.

6.1 Molecular mechanisms of apoptosis

Cells may be induced to undergo apoptosis by extracellular death factors or by internal physical/chemical insults such as DNA damage or oxidative stress. Subsequently two non-exclusive molecular pathways, the

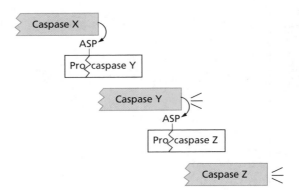

Figure 6.1 A caspase cascade

extrinsic and the intrinsic respectively, may be activated. Caspases, specific proteases that cleave intracellular proteins at aspartate residues ◎ (one of the 20 amino acids) like molecular scissors, are central to both apoptotic pathways. The term "caspases" arose from three of their enzyme characteristics: they are cysteine-rich aspartate proteases. The proteases of this family are synthesized as inactive enzymes called procaspases that need to be cleaved at aspartate residues in order to be activated. Although for the most part procaspases are considered inactive, procaspases possess some activity; they possess about 2% of the proteolytic activity of fully activated caspases. This may seem insignificant at the moment but, as we will see below, it is an important feature for some pathways of caspase activation. Moreover, since caspases cleave at aspartate residues and procaspases are activated by cleavage at aspartate residues, caspases participate in a cascade of activation whereby one caspase can activate another caspase in a chain reaction (Figure 6.1). This mechanism, whereby caspases activate procaspases, leads to amplification of an apoptotic signal since only a few initially activated caspase molecules can produce the rapid and complete conversion of a pool of procaspases. Let us examine both the extrinsic and intrinsic apoptotic pathways below.

The extrinsic pathway: mediated by membrane death-receptors

The extrinsic pathway for triggering cell death (Figure 6.2) share some common features with pathways involved in triggering cell growth (Chapter 4). A death factor such as Fas ligand or tumour necrosis factor (TNF) is received by a transmembrane death receptor such as Fas receptor or TNF receptor, respectively. TNF is a soluble factor while Fas ligand is bound to the plasma membrane of neighboring cells. When ligands bind to the death receptors, the receptors undergo a conformational

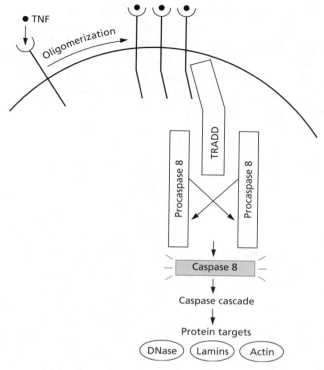

Figure 6.2 The extrinsic pathway of apoptosis

change and oligomerize in order to transduce the signal into the cell. The conformational change exposes so-called death domains that are located on the receptors' cytoplasmic tail and enable intracellular adaptor proteins such as FADD (Fas-associated death domain protein) and TRADD (TNF receptor-associated death domain protein) to bind via their death domains. See Pause and Think.

The adaptors function to transduce the death signal from the receptor to caspases. The adaptors recruit several molecules of procaspase-8, also known as an initiator caspase since it is the first link between the receptor and the apoptotic proteases. Procaspase-8 molecules, now in close proximity to each other, become activated by self-cleavage due to their low enzymatic activity and initiate a cascade of caspase activation: one activated caspase cleaves and activates another. The cascade ultimately results in the cleavage of specific target substrates.

The proteolysis of the target proteins facilitate the breakdown of the cell; they include nuclear lamins allowing for nuclear shrinkage, cytoskeletal proteins such as actin and intermediate filaments for rearranging cell structure, specific kinases for cell signaling, and other enzymes such as caspase-activated DNase for the cleavage of chromatin. The caspase-activated DNase cuts DNA between nucleosomes and generates

PAUSE AND THINK

What do you think is the function of a death domain? A death domain is part of a protein that allows for specific protein–protein interactions to occur and is analogous to the SH2 domain characteristic of growth factor signal transduction pathways.

a DNA ladder (corresponding to multiples of 180 bp—the distance between nucleosomes), that can be detected experimentally and used by scientists as a molecular marker of apoptosis. The TUNNEL technique, described below, is another procedure used by scientists to detect apoptosis. Caspases also cleave the tumor suppressor protein, pRB, and this cleavage results in the degradation of pRB protein. This event is required for apoptosis induced by TNF and points to a role for pRB in the inhibition of apoptosis.

Self test   Close this book and try to redraw Figure.6.2. Check your answer. Correct your work. Close the book once more and try again.

Analysis of apoptosis by the TUNNEL technique

As mentioned above, specific fragmentation of DNA at internucleosomal sites is characteristic of apoptosis. The apoptotic endonucleases generate free 3′ OH groups at the ends of the DNA fragments that can be experimentally end-labeled using tagged nucleotides. The enzyme terminal deoxynucleotidyl transferase (TdT) catalyzes the addition of labeled deoxynucleotides to the 3′ OH ends of the many DNA fragments within an apoptotic nucleus. Biotin is a common label that can be detected using diaminobenzidine and a streptavidin-horseradish peroxidase conjugate to generate a color reaction at the site of the DNA ends.

The intrinsic pathway: mediated by the mitochondria

The intrinsic pathway of apoptosis (Figure 6.3) does not depend on external stimuli. Internal stimuli, such as DNA damage and oxidative stress, induce apoptosis through the Bcl-2 family of proteins that act at the outer mitochondrial membrane. The Bcl-2 family consists of approximately 20 members, all of which contain at least one Bcl-2 homology (BH) domain that mediates protein–protein interactions. Most family members share three or four BH domains. There are two groups within the Bcl-2 family that have opposing functions: one group of Bcl-2 proteins inhibit apoptosis and another group promotes apoptosis (Table 6.1). Within the group of pro-apoptotic molecules is a subset referred to as the BH3-only proteins because they only share one BH domain, BH3. The Bcl-2 family of proteins are thought to act either by forming channels or blocking channels in the outer mitochondrial membrane to regulate the release of important molecular apoptotic mediators from the mitochondria. The members of this family can associate by protein–protein interactions and it appears that the balance of factor function determines the outcome. For example, if the activity of the

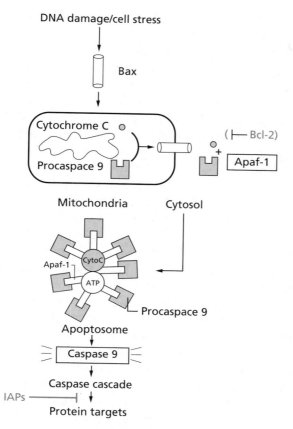

Figure 6.3 The intrinsic pathway of apoptosis

Table 6.1 Members of the Bcl-2 family

Anti-apoptotic members	Pro-apoptotic members	BH3 only members
Bcl-2	Bax	Bad
Bcl-x$_L$	Bok/Mtd	Bik/Nbk/Blk
Bcl-w	Bcl-x$_s$	Bid
A1	Bak	Hrk/DP5
Mcl-1	Bcl-G$_L$	Bim/Bod
Boo		Bmf
		Noxa
		Puma/Bbc3

pro-apoptotic factors is high due to low inhibition from anti-apoptotic factors, apoptosis is triggered.

The intermembrane space between the two mitochondrial membranes acts as a supply cabinet for apoptotic mediators. The pro-apoptotic Bcl-2 members regulate the release of the apoptotic mediators from this mitochondrial compartment. Upon activation by an apoptotic signal, Bcl-2 members such as Bax and Bak undergo a conformational change and insert into the outer mitochondrial membrane. Their insertion increases the permeability of the outer mitochondrial membrane by forming and/or regulating membrane channels and allows the release of apoptotic mediators. As shown in Figure 6.3, cytochrome C, which also functions in the electron transport chain of oxidation, and procaspase-9 are released into the cytoplasm and assemble into a complex called an apoptosome, along with ATP and Apaf-1. The binding of Apaf-1 to cytochrome c facilitates the recruitment of procaspase-9 via protein domains, called CARD domains, present on both Apaf-1 and procaspase-9. Apaf-1 is a protein cofactor that is required for procaspase-9 activation. Upon activation, caspase-9 begins another caspase cascade by activating caspase-3. Other factors, such as inhibitors of apoptosis proteins (IAP) play a role modulating the process. The x-chromosome linked member, XIAP, is one member of the IAP family that directly binds to and inhibits the activity of caspase-9, caspase-3, and caspase-7 after they have been processed.

Note that there is cross-talk between the extrinsic and intrinsic pathways and the two converge at the activation of downstream caspases. For example, caspase-8 can proteolytically cleave and activate the Bid, a pro-apoptotic Bcl-2 family member. Bid can then stimulate the intrinsic pathway of apoptosis by facilitating the release of cytochrome c from the mitochondria and inducing the subsequent activation of downstream caspases.

Self test   Study Table 6.1. It is crucial to remember that Bcl-2 blocks apoptosis and this fact will help you to remember the role of the other Bcl-2 family members. Close this book and try to redraw Figure 6.3. Check your answer. Correct your work. Close the book once more and try again.

---

p53 and apoptosis

As we saw in Chapter 5, the tumor suppressor protein, p53, accomplishes its role as the guardian of the genome, in part, by inducing apoptosis in response to DNA damage and cellular stress. It does this by both transcription-dependent and -independent means. As a transcription factor, p53 induces the expression of genes that code for death receptors and pro-apoptotic members of the Bcl-2 family (see Table 5.2).

Examples include *Fas* receptor, *Bax*, and *Bak*. These genes contain a consensus p53 binding site in their promoter regions. p53 can also repress the expression of anti-apoptotic factors, such as Bcl-2 and Bcl-x and IAPs. Recent *in vivo* evidence demonstrates a member of the Bcl-2 family called *PUMA* (p53 upregulated modulator of apoptosis), a target of p53, is essential for apoptosis induced by p53. Apoptosis induced by DNA-damaging drugs, irradiation, oncogenic activation, and cell stress was blocked in *PUMA* gene knockout mice. p53 can exert transcription-independent regulation of apoptosis. This has been demonstrated by the induction of apoptosis with p53 mutants incapable of transcription; these p53 mutants included one that lacked the DNA-binding domain disabling interaction with p53 target genes, one without a nuclear localization signal preventing p53 from reaching the target genes in the nucleus, and one mutated in the p53 transactivation domain preventing transcriptional activation function. It was also demonstrated that UV-induced apoptosis could be triggered by wild-type p53 strictly from the cytoplasm in cells that were treated with wheatgerm agglutinin, a nuclear import inhibitor. The mechanism of p53 transcription-independent apoptosis involves p53 activation of Bax in the cytoplasm and subsequent cytochrome c release and caspase activation (Chipuk *et al.*, 2004). Evidence also supports the role of p53 in releasing pro-apoptotic proteins (e.g. Bid) from sequestration by anti-apoptotic proteins (e.g. Bcl-x_L), altering the net functional balance of the Bcl-2 family of proteins. In summary, p53 functions in both the nucleus and the cytoplasm by transcription-dependent and independent means.

6.2 Apoptosis and cancer

Evasion of apoptosis is one of the six hallmarks of cancer (Figure 1.1). Tumor cells possess many signals, such as DNA damage and oncogene activation, that normally induce apoptosis. Through tumor suppression pathways, most cells that acquire carcinogenic characteristics are eliminated by apoptosis. However, tumor cells that acquire mutations that allow them to escape from the apoptotic response survive and proliferate. The avoidance of apoptosis permits further accumulation of mutations. This draws our attention to a difference that develops between tumor cells and normal cells (Figure 6.4): tumor cells are more close to triggering an apoptotic response if the apoptotic pathway is somehow repaired compared with normal cells (a concept that has been considered for the design for new cancer therapeutics by targeting p53; Chapter 5). Oxidative stress and oncogene activation are upstream activators of apoptosis. Normal cells do not possess as many apoptotic-inducing signals.

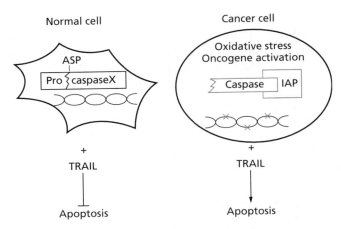

Figure 6.4 Apoptotic responses in normal cells vs. cancer cells

In addition, evidence suggests that there is a fundamental difference in the state of caspase activation between cancer cells and normal cells (Figure 6.4): cancer cells contain **activated** caspases that are inhibited by upregulated IAPs but normal cells require proteolytic cleavage of fairly inactive procaspases. Therefore, cancer cells are "closer" to triggering an apoptotic response once a death signal is received compared with a normal cell. In support of this, processed caspase-3 has been identified in tumor cells. Tumors cell share this characteristic with the normal *Drosophilia* apoptotic process whereby activated caspases are inhibited by IAPs in normal cells and induction of apoptosis requires release from IAP inhibition. In summary: apoptotic signals stimulate procaspase processing in normal cells, while apoptotic signals stimulate the release of IAP inhibition of processed caspases in cancer cells.

A subfamily of TNF receptors, called TRAIL receptors, has been found to elicit a differential sensitivity to apoptosis between normal cells and cancer cells. Its ligand, TRAIL (TNF-related apoptosis-inducing ligand), induces apoptosis in many cancer cells (regardless of the *p53* gene profile) but not in most normal cells. This subfamily signals apoptosis in a similar manner to TNF receptors recruiting adaptors (e.g. FADD) and an initiator caspase, caspase-8, to the membrane (see Figure 6.2). Since TRAIL and its receptors are expressed in most organs, it is hypothesized that the addition or loss of regulatory molecules determines whether apoptosis will be induced in particular cell types.

These properties create an opportunity for designing drugs that target mutations occurring in the apoptotic pathways and suggest that attempts to restore apoptotic activity will not affect normal cells. Such drugs are likely to have few side effects. Before discussing apoptotic therapies, we will first examine a sample of mutations in the apoptosis pathways that have been identified in cancers.

Mutations that affect the extrinsic pathway

Mutations in death receptor genes, such as those encoding the Fas receptor and TRAIL receptor, occur in some cancers. Fas ligand and receptor are induced by UV light. Signaling through Fas receptors induces the development of sunburn in response to UV and is an important defense against skin cancers (Guzman *et al.*, 2003). Somatic mutations in Fas receptors have been reported in melanomas and squamous cell carcinomas.

Suppression of caspase gene expression in the development of particular cancers, such as small-cell lung carcinoma and neuroblastomas, has also been demonstrated. Remember that caspase-8 is the first caspase activated by the death receptors (Figure 6.2) and maintains a top position in the initiation of the caspase cascade. Loss of caspase-8 expression observed in cancers is due to both epigenetic and mutational alterations; hypermethylation of the caspase-8 promoter, deletions, and missense mutations have been identified. Caspase-8 deficiency is particularly characteristic of neuroblastomas and small-cell lung cancer. Methylation of the caspase gene promoter was found to be the mechanism of loss of expression in a majority of cell lines examined. Also, a deletion of leucine 62 was identified in human vulval squamous carcinoma cells and this mutation blocks the interaction of caspase-8 with the adaptor FADD thus abolishing its link with the death signal and receptor.

Mutations that affect regulators of the intrinsic pathway

Alterations of the intrinsic pathway of apoptosis are much more common than alterations in the extrinsic pathway during carcinogenesis. Mutations that bypass the transduction of apoptotic signals that are triggered by DNA damage, a dominant characteristic of cancer cells, are favored by natural selection. A major contribution to intrinsic pathway alterations occurs through mutations that affect the p53 pathway. Mutations in the *p53* gene itself are the most frequently found mutations in cancer cells, and current estimations may be underestimated since many earlier studies restricted their analysis to exons 5–9 only, rather than the full length of the gene. The *p53* mutations provide the cancer cells with a survival advantage by disrupting apoptosis. Abnormal methylation and loss of heterozygosity for *p73*, a *p53* family member capable of inducing apoptosis, is frequent in lymphomas. Also, the *p73* gene undergoes alternative splicing to generate several RNAs, including N-terminally deleted variants which can act as inhibitors of *p53*. These variants are over-expressed in many cancers. In addition, mutations in genes involved in the upstream regulation of *p53* (e.g. *ATM* and *Chk2*; see Chapter 5, Figure 5.7) and also in p53s downstream targets have also been identified in human tumors. Mutations in molecular components that affect

MDM2, the major regulator of p53 activity, are common in tumors that maintain wild-type p53 alleles.

Bcl-2, the first member of the Bcl-2 family of proteins to be discovered, was initially identified from a chromosomal translocation, t(14:18), in B-cell lymphomas, hence the name *bcl*. In t(14:18), the *Bcl-2* gene is translocated to a position juxtaposed to the immunoglobulin heavy chain enhancer. As a consequence of its relocation next to a strong promoter, oncogenic activation of the *Bcl-2* gene occurs. Overexpression of the anti-apoptotic protein, Bcl-2, leads to insufficient apoptotic turnover and B-cell accumulation. This translocation is not only found in most cases of follicular B-cell lymphomas but is also found in other types of cancer such as gastric, lung, and prostate. Aberrant expression of most of the genes in the Bcl-2 family is linked to carcinogenesis. All anti-apoptotic members of the Bcl-2 family may function as oncogenes and pro-apoptotic members act as tumor suppressor genes. Mutations in genes that code for pro-apoptotic proteins, such as deletions in the *bak* and *bid* gene, are characteristic of some tumors. *Bax* is mutated in over 50% of a specific class of colon tumors. Remember that p53 regulates many genes of the Bcl-2 family. Thus, many mutations in *p53* that are common in tumors also affect the transcriptional regulation of its target genes, including those of the Bcl-2 family.

Molecules involved in events downstream of the release of mitochondrial apoptotic factors also play a role in tumorigenesis. The gene encoding Apaf-1, the coactivator of caspase-9 upon its release from the mitochondria, is mutated and transcriptionally repressed in metastatic melanoma. Note that epigenetic inactivation, in addition to mutation, plays a role in the inactivation of the apoptotic pathway.

The induction of inhibitors of apoptosis also plays a role in carcinogenesis. XIAP is induced in many types of cancer including leukemias, lung cancer and prostate cancer. Since XIAP acts to suppress caspases 9, 3, and 7, it affects downstream caspases that are common to both the extrinsic and intrinsic pathways.

Alternative death pathways

The observation that many apoptotic stimuli do not require caspases has led to studies of alternative death pathways. Similar to apoptosis, these alternative death pathways depend on internal signals, and are therefore defined as types of programmed cell death.

The details of the molecular events involved in caspase-independent cell death are not complete. However, it is known that they also utilize proteases and also facilitate permeabilization of the mitochondrial membrane. Alternative proteases such as calpains, cathepsins, and serine proteases cleave target proteins to bring about morphological changes characteristic of programmed cell death. Calpains, like caspases, are found as inactive zymogens in the cytoplasm. Cathepsins become activated in lysosomes before being translocated into the cytoplasm and/or nucleus. Apoptosis-inducing factor (AIF) is one molecular player released from the mitochondrial intermembrane space that induces caspase-independent DNA degradation.

The abnormal expression of molecules involved in alternative death pathways is observed in tumor cells. For example, mutations in genes that encode the tumor suppressor proteins, Bin1 and promyelocytic leukemia (PML) protein, that induce alternative death pathways are found in human cancers. Bin activates a caspase-independent pathway that is blocked by a serine protease inhibitor. As more is learned about the molecular players of alternative death pathways, new potential drug targets will be uncovered. In fact, several drugs in clinical trials (e.g. EB1089/seocalcitol, a vitamin D analogue) induce calpain-dependent and caspase-independent cell death.

6.3 Apoptosis and chemotherapy

Disruption of the apoptotic pathway has important effects on the clinical outcome of chemotherapy. In order for chemotherapy to be successful, cells must be capable of undergoing apoptosis. Chemotherapeutic agents act primarily by inducing DNA damage and this damage consequently triggers the intrinsic apoptotic pathway. Drugs with varying structures indirectly elicit the same morphological changes typical of apoptosis. However, remember that one of the hallmarks of cancer cells is that they evade apoptosis. Many tumors have defective apoptotic pathways and are inherently resistant to chemotherapies, regardless of whether or not they have been previously exposed to the drugs. This type of resistance contrasts the classical acquired mechanisms that are associated with drug accumulation and drug stability such as the P-glycoprotein pump (Chapter 2). Since chemotherapy resistance is a major clinical problem, elucidating the role of apoptosis in drug responses is important for future therapeutic strategies.

Drug resistance can arise through mutations in genes that code for molecular regulators of apoptosis. These mutations serve to uncouple drug-induced damage from the activation of apoptosis. As mentioned above, mutations in the p53 pathway are common in cancer cells and greatly contribute to the inherent drug resistance observed for many cancers. Cells engineered to have a *p53* knockout are resistant to drug-induced apoptosis. Yet, some "gain of function" *p53* mutations may confer resistance to specific chemotherapies. One p53 mutant induces the expression of the *dUTPase* gene and results in resistance to 5-fluorouracil.

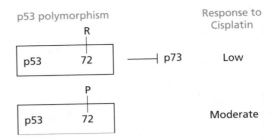

Figure 6.5 p53 polymorphisms affect drug response

Therefore both "loss of function" and some "gain of function" mutations can give rise to resistance. On the other hand, one mutant p53 sensitizes some cells to taxanes.

The p53 family member, p73, can also mediate an apoptotic response to DNA damage. Many chemotherapeutic agents, such as cisplatin and doxorubicin induce *p73*, implicating that p73 also plays a role in drug-induced apoptosis. Experimental evidence from siRNA studies (see Chapter 1) has shown that inhibition of p73 function causes an increase in drug resistance. In several chapters of this text, we will see that gene polymorphisms affect certain biological responses; here, one example is illustrated by a p53 polymorphism that influences the response rate to cisplatin therapy (Bergamaschi *et al.*, 2003). Patients with the 72R (arginine at amino acid position 72) respond less well than patients with 72P (proline at amino acid position72) because 72R mutants inhibit p73 (Figure 6.5). The 72R polymorphism is more common in cancer cells.

The upregulation of the anti-apoptotic members of the Bcl-2 family and the downregulation of the pro-apoptotic members of the Bcl-2 family in tumors is associated with an increased resistance to chemotherapies. For example, loss of Bax, a pro-apoptotic protein, increases drug resistance in human colorectal cancer cells to the antimetabolite 5-fluorouracil and nonsteroidal anti-inflammatory drugs (NSAIDS) used as chemopreventative agents (Zhang *et al.*, 2000). Over expression of Bcl-2 in metastatic tumors may contribute to the fact that they are notoriously chemoresistant.

Overall, these observations point to an important clinical implication: the genotype of a tumor, especially with respect to the *p53* and *Bcl-2* gene families, is an important factor that influences the effectiveness of therapy.

There is another important implication of treating cells that have nonfunctional apoptotic pathways with chemotherapy. The lack of an apoptotic effect in response to extensive DNA damage caused by these genotoxic drugs provides an opportunity for the accumulation of mutations. Consequently the risk of carcinogenesis increases. Indeed therapy-related leukemia, whereby a new cancer arises after chemotherapy administration, is a clinical problem. Therapy-related leukemias have relatively short latency times. Specific cytogenetic aberrations are associated with different chemotherapeutic agents; chromosomal deletions of #5 and/or #7 are characteristic of alkylating drugs. (For a recent case study see Griesinger, 2004.)

◎ Therapeutic strategies

6.4 Apoptotic drugs

The ability to trigger apoptosis in tumor cells is an important strategic design for cancer therapeutics. This is supported by the fact that many successful conventional chemotherapies work by triggering apoptosis, albeit indirectly. Using the knowledge of the molecular players in the apoptotic pathways enable us to design direct apoptotic inducers. Alternatively, endogenous inhibitors can be blocked. This approach bypasses the need for a drug to be mutagenic and avoids therapy-related leukemias. In addition, the induction of apoptotic factors in normal cells should have little effect on these cells since they are not poised to trigger apoptosis to the same degree as tumor cells. See Pause and Think.

Below is a description of strategies targeted against caspases, the Bcl-2 family, and TRAIL.

Direct and indirect activation of caspases

Selective activation of caspases is the most obvious apoptotic target. However, because they comprise a large family of over 12 members and are in every cell type, selectivity has been a problem thus far. Yet, Maxim Pharmaceuticals has several small molecule caspase activators (e.g. MX-2060) that are in early stages of development. Procaspase-3, the proenzyme of a key effector caspase, is inhibited by an intramolecular interaction facilitated by three aspartate residues, called the "safety-catch". Screens are being pursued for small molecules that are able to interfere with this intramolecular inhibitory conformation of procaspase-3. Time is needed to see if caspases are successful cancer therapeutic targets.

However, the activation of caspases by indirect methods is promising. Since endogenous caspase inhibitor, XIAP, is overexpressed in many cancers it is a good molecular target for new cancer therapeutics. A synthetic chemical screen aimed at inhibiting XIAP activity identified a class of polyphenylureas that directly relieves the inhibition of caspase-3

PAUSE AND THINK

What molecules would you target if asked to design an apoptotic drug?

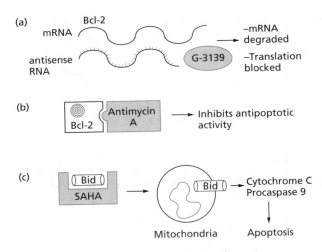

Figure 6.6 Drug strategies that target the Bcl-2 family of proteins

and caspase-7 but not caspase-9 (Schimmer *et al.*, 2004). This is feasible because one domain of XIAP directly blocks the active site of caspase-3 and -7 while another distinct domain inhibits caspase-9. The small molecule inhibitors bind to the XIAP domain known to block the active site of caspase-3 and -7, the downstream caspases. Furthermore, these compounds induced apoptosis in a range of tumor cell lines and showed anti-tumor activity in animal tumor models. Little toxicity was observed for normal cells. This was the first demonstration that relief of caspase inhibition can induce tumor cell apoptosis directly. Interestingly, other XIAP inhibitors (e.g. Smac peptides) were less successful due to their selective relief of caspase-9 inhibition only, thereby leaving downstream caspases-3 and -7, available for inhibition.

Regulation of the Bcl-2 family of proteins

The Bcl-2 family is another target for the design of apoptotic drugs. Three main strategies have been used and these are illustrated in Figure 6.6: a) antisense RNA, and b) small molecules to inhibit protein function and protein–protein interactions of anti-apoptotic molecules, and c) drugs that induce the activity of pro-apoptotic molecules. Since Bcl-2 is overexpressed in a broad range of tumors, inhibition of its expression by antisense is one strategy that has been employed to create a new cancer drug. G-3139 (Genasense) is an 18-mer modified antisense oligonucleotide that is complementary to the first six codons downstream of the translational start site on the Bcl-2 mRNA. Upon *in vivo* hybridization, translation is inhibited and the mRNA is degraded. This alters the balance of pro- and anti-apoptotic factors in favor of apoptosis. Anti-tumor effects have been demonstrated in early clinical trials and G-3139 is currently in phase III trials (Banerjee, 2001). Small molecule inhibitors that bind to Bcl-2/Bcl-x$_L$ and interfere with protein–protein interactions with pro-apoptotic molecules in order to induce apoptosis have been identified. However, this approach has not yet reached clinical trials. Suberoylanilide hydroxamic acid (SAHA) works by an opposing mechanism; instead of inhibiting an anti-apoptotic factor, it promotes the activity of a pro-apoptotic factor. SAHA is a HDAC inhibitor and therefore acts to induce the expression of epigenetically suppressed genes. SAHA induces the expression of the pro-apoptotic protein, BID, which translocates into the mitochondria and results in the release of apoptotic factors such as cytochrome C. Phase II trials have demonstrated anti-tumor activity.

Targeting TRAIL and its receptor

The differential activity of TRAIL and its receptor in normal cells vs. cancer cells, suggests that they are good molecular targets for apoptotic therapies. Approximately 80% of cancer cell lines are sensitive to TRAIL ligand and can be induced to undergo apoptosis. See Pause and Think.

The administration of recombinant human TRAIL ligand has exhibited promising anti-tumor activity in animal models but Phase I trials have yet to be initiated. Caution is being exercised because of the hepatic toxicity known to result from related Fas and TNF-α ligand administration and also because of one study that reported a TRAIL-induced apoptotic response in human hepatocytes in culture even though toxicity was not observed in mice and nonhuman primates. The variability of the effects among different species reminds us that there are limitations of animal studies for predicting toxicity of cancer therapies in humans. Also, the critical eye will notice that different structural forms of recombinant TRAIL were used in different studies due to variation in the preparation of the recombinant protein and thus may be a source of the toxicity. Further along is the testing of a TRAIL receptor agonistic monoclonal antibody that recognizes the receptor's extracellular domain. Patients have been enrolled in phase I trials to evaluate drug pharmacology and toxicity.

> **PAUSE AND THINK**
>
> What strategy would you use to induce apoptosis through activation of the TRAIL receptor?

■ CHAPTER HIGHLIGHTS—REFRESH YOUR MEMORY

- Apoptosis is an important tumor suppression mechanism.

- Caspases, aspartate proteases, are main molecular players during apoptosis.

- Apoptosis can be triggered by extracellular death signals or internal stimuli that act via an extrinsic and intrinsic pathway, respectively.

- TNF/TNFR and FAS/FASR signaling are paradigms of the extrinsic pathway.

- The mitochondria stores apoptotic molecules involved in the intrinsic pathway.
- The Bcl-2 family regulate the permeability of the outer mitochondrial membrane.
- p53 induces apoptosis by both transcription-dependent and -independent means.
- Evasion of apoptosis is a hallmark of cancer cells.
- A tumor cell is closer to eliciting an apoptotic response than a normal cell if the apoptotic pathway is functional.
- Caspase activity is regulated differently in normal cells and tumor cells: normal cells require procaspase processing while tumor cells require release of processed caspases from IAPs.
- Alterations in the p53 and Bcl-2 related pathways play a major role during carcinogenesis.
- Chemotherapies act indirectly via DNA damage to induce apoptosis.
- Tumors with defective apoptotic pathways are resistant to chemotherapies.
- Mutations in apoptotic proteins enable cancer cells to both survive and become drug resistance.
- Treatment with chemotherapies can cause therapy-related leukemia.
- Apoptotic drugs aim to trigger apoptosis directly and do not require genotoxic activity.

■ ACTIVITY

(i)   It has been reported that the transcription factor NFκB blocks apoptosis in several cell types. After consulting the scientific literature (Hint: begin with: Li, X. and Stark, G.R. (2002) NFκB-dependent signaling pathways. *Exp. Hematol.* **30**: 285–296) draw a pathway diagram illustrating the molecular components involved in the inhibition of apoptosis by NFκB. Have mutations in this pathway been identified in cancer cells?

■ FURTHER READING

Danial, N.N. and Korsmeyer, S.J. (2004) Cell death: critical control points. *Cell* **116**:205–219.

Gasco, M. and Crook, T. (2003) P53 family members and chemoresistance in cancer: what we know and what we need to know. *Drug Resist. Updates* **6**:323–328.

Hickman, J.A. (2002) Apoptosis and tumourigenesis. *Curr. Opin. Genet. Dev.* **12**:67–72.

Hengartner, M.O. (2000) The biochemistry of apoptosis. *Nature* **407**:770–776.

Hu, W. and Kavanagh, J.J. (2003) Anticancer therapy targeting the apoptotic pathway. *Lancet Oncol.* **4**:721–729.

Johnstone, R.W., Ruefli, A.A. and Lowe, S.W. (2002) Apoptosis: A link between cancer genetics and chemotherapy. *Cell* **108**:153–164.

Kirkin, V., Joos, S. and Zornig, M. (2004) The role of Bcl-2 family members in tumorigenesis. *Biochim. Biophys. Acta* **164**:229–249.

Los, M., Burek, J.C., Stroh, C., Benedyk, K., Hug, H. and Mackiewicz, A. (2003) Anticancer drugs of tomorrow: apoptotic pathways as targets for drug design. *DDT* **8**:67–77.

Makin, G. and Dive, C. (2001) Apoptosis and cancer chemotherapy. *Trends Cell Biol.* **11**:S22–S26.

Mathiasen, I.S. and Jaattela, M. (2002) Triggering caspase-independent cell death to combat cancer. *Trends Mol. Med.* **8**:212–220.

Vousden, K.H. and Lu, X. (2002) Live or let die: the cell's response to p53. *Nature Rev. Cancer* **2**:594–604.

Zhivotovsky, B. and Orrenius, S. (2003) Defects in the apoptotic machinery of cancer cells: role in drug resistance. *Sem. Cancer. Biol.* **13**:125–134.

■ WEB SITES

www.genta.com

Maxim Pharmaceuticals www.maxim.com

■ SELECTED SPECIAL TOPICS

Banerjee, D. (2001) Genasense (Genta Inc.). *Curr. Opin. Invest. Drugs* **2**:574–580.

Bergamaschi, D., Gasco, M., Hiller, L., Sullivan, A., Syed, N., Trigiante, G., Yulug, I., Merlano, M., Numico, G., Comino, A., Attard, M., Reelfs, O., Gusterson, B., Bell, A.K., Heath, V., Tavassoli, M., Farrell, P.j., Smith, P., Lu, X. and Crook, T. (2003) p53 polymorphism influences response in cancer chemotherapy via modulation of p73-dependent apoptosis. *Cancer Cell* **3**:387–402.

Chipuk, J.E., Kuwana, T., Bouchier-Hayes, L., Droin, N.M., Newmeyer, D.D., Schuler, M. and Green, D.R. (2004) Direct activation of Bax by p53 mediates mitochondrial membrane permeabilization and apoptosis. *Science* **303**:1010–1014.

Griesinger, F., Metz, M., Trumper, L., Schulz, T. and Haase, D. (2004) Secondary leukemia after cure for locally advanced NSCLC: alkylating type second leukemia after induction therapy with docetaxel and carboplatin for NSCLC IIIB. *Lung Cancer* **44**:261–265.

Guzman, E., Langowski, J.L. and Owen-Schaub, L. (2003) Mad dogs, Englishman and apoptosis: The role of cell death in UV-induced skin cancer. *Apoptosis* **8**: 315–325.

Schimmer, A.D., Welsh, K., Pinilla, C., Wang, Z., Krajewska, M., Bonneau, M.-J., Pedersen, I.M., Kitada, S., Scott, F.L., Bailly-Maitre, B., Glinsky, G., Scudiero, D., Sausville, E., Salvesen, G., Nefzi, A., Ostresh, J.M., Houghten, R.A. and Reed, J.C. (2004) Small-molecule antagonists of apoptosis suppressor XIAP exhibit broad anti-tumor activity. *Cancer Cell* **5**:25–35.

Zhang, L., Yu, J., Park, B.H., Kinzler, K.W. and Vogelstein, B. (2000) Role of BAX in the apoptotic response to anticancer agents. *Science* **290**:989–992.

Zhivotovsky, B. and Orrenius, S. (2003) Defects in the apoptotic machinery of cancer cells: role in drug resistance. *Sem. Cancer Biol.* **13**:125–134.

Stem cells and differentiation

Introduction

As described in Chapter 1, the balance between cell growth, differentiation, and apoptosis affects the net numbers of cells in the body and aberrant regulation of these processes can give rise to tumors. In this chapter we will describe the characteristics of cells at different degrees of differentiation and discuss the relationship of these characteristics to those of cancer cells. We will also investigate the molecular mechanisms that underlie the regulation of differentiation and examine specific mutations in differentiation pathways that can lead to cancer. Lastly, new cancer therapeutics designed to target different aspects of differentiation pathways are presented. Let us begin with an overview of the process of differentiation during development and in the adult.

We seldom reflect upon our own ontogeny, or individual development. The processes involved in the development of a complete person from a formless fertilized egg are almost magical. Hundreds of specialized cell types must form from the fertilized egg and its unspecialized progeny cells called embryonic stem cells, that reside in the inner cell mass. The process whereby cells become specialized to perform a particular function is called differentiation and relies on the regulation of a particular subset of genes that define a certain cell-type. All cells in the body (except red blood cells) contain a full complement of genes of the human genome but it is the expression of a subset of genes that makes one cell type different from another: for example a brain cell expresses different genes compared with a liver cell. Lineage-specific transcription factors responsible for turning on cell-type specific genes are important in this process. During our development, different cell types are organized into varying tissues by pattern formation; although the same cell types are present in an arm and a leg, the morphology, or form of the structures, differs. Regulated gene expression is also important for patterning during development.

In addition to embryonic stem cells, there are also stem cells in the adult that are involved in the regeneration of tissues during the lifetime of the individual. In fact, stem cells are believed to be present in all tissues. Some stem cells are continually active to replace cells as they mature and die off. For example, skin stem cells self-renew and regenerate skin that is lost to the environment. Other stem cells remain dormant until a physiological signal is received. Breast stem cells strongly respond to pregnancy hormones and to a lesser extent to hormones within the monthly cycle. Adult hematopoietic stem cells, stem cells that give rise to the blood, self-renew and differentiate to sustain the different types of blood cells over the lifetime of the individual. It has recently been demonstrated that hematopoietic stem cells show differentiation plasticity—that is they can give rise to nonhaematopoietic cells. See Pause and Think.

But perhaps this should not be too surprising. Recent experiments of cloning have demonstrated that a nucleus from a differentiated cell can be reprogrammed to direct the development of another individual. The cloning of Dolly the sheep, from a mammary cell nucleus, is a notable demonstration that the pattern of gene expression of a differentiated cell is not permanently fixed.

The process of differentiation is fueled by a source of stem cells in both the embryo and the adult. Stem cells self-renew while at the same time give rise to cells that are more committed to differentiate along a particular cell lineage. Differentiated cells are associated with withdrawal from the cell cycle. A block in differentiation results in a higher net number of cells and therefore is an important feature for tumor formation in some cancers. Also, despite all the cells from a particular tumor being of clonal origin (Section 1.2), the tumor contains a mix of cells that have different genetic and physical characteristics. Restated in genetics terminology, a tumor is a mass of genotypically and phenotypically heterogeneous cells despite all the cells being of clonal origin. This heterogeneity may reflect aberrant differentiation and development in addition to accumulation of different mutations. Indeed, the differentiation status of a tumor may be indicative of patient prognosis. The most malignant tumors often show the least amount of differentiation markers. Although tumors may exhibit features of differentiation (e.g. teeth in teratomas), the normal pattern formation of cells that underlies the morphology of a normal organ is abandoned. We will examine two features of differentiation pathways that have implications to carcinogenesis: first, the characteristics of stem cells, which are the precursors of differentiated cells, and secondly, the role of lineage-specific transcription factors that act as master switches for sets of genes during the differentiation process.

7.1 Stem cells and cancer

Two defining features of stem cells are their ability to self-renew and their ability to give rise to differentiated cell types of one or more cell lineages. Upon cell division, one daughter cell maintains the characteristics of a stem cell including the ability to self-renew and the other daughter cell shows characteristics of commitment towards differentiation (Figure 7.1). The feature of self-renewal is shared with tumor cells. This common feature has led to two proposals for the relevance of stem cell biology to carcinogenesis. The first proposal is that self-renewal provides increased opportunities for carcinogenic changes to occur. The second proposal suggests that altered regulation of self-renewal directly underlies carcinogenesis. We will explore each of these below where we will find that the concepts from both proposals will intertwine. Note that research into this topic is relatively in its infancy and these proposals require further investigation and substantiation.

Self-renewal provides an extended window of time for mutation

Stem cells are long-lived targets for chance mutations compared with many differentiated cells that die within days or months. The accumulation of mutations necessary for carcinogenesis is more likely to occur in stem cells that self-renew over the life-time of an individual rather than in mature cells that exit the cell cycle and/or undergo apoptosis after a brief period. This concept supports the proposal the tumors are likely to arise from stem cells. Because restricted progenitors cells self-renew for shorter periods of time they are less likely than stem cells to become oncogenic. Let us look at the process of skin carcinogenesis. Skin is a tissue with a clearly defined hierarchical organization of differentiation. The differentiation pathway begins in the basal layers and leads to the formation of the dead, cornified outer layer of the skin. Epithelial cells of the skin have

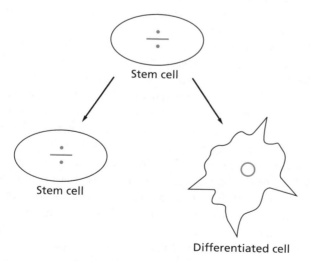

Figure 7.1 Features of stem cells

a turnover rate of 60 days in humans. However, malignant transformation involving the accumulation of specific mutations takes 18 months or more, so the 60-day life-span of a differentiating cell is not long enough to allow a sufficient number of mutations to accumulate. Self-renewal is a quality of stem cells that allows for the accumulation of transforming mutations since an individual mutation may be passed on to daughter cells which are themselves susceptible to additional mutations and can pass on accumulated mutations to future progeny cells. This rationale suggests that the accumulation of mutations required for the initiation of cancer is likely to occur in the normal stem cell compartment.

Deregulation of self-renewal

Stem cells must maintain a balance between self-renewal and differentiation. One proposal for the relevance of stem cell biology to carcinogenesis is that the loss of this balance by stem cells can lead to unregulated self-renewal, a hallmark of cancer. Therefore, tumor cells may arise from stem cells. Alternatively, differentiated cells may acquire a mutation that reactivates a self-renewal program. Perhaps in a manner similar to that in which viruses 'hijack' the cell's machinery for replication, tumor cells 'hijack' the machinery of self-renewal for oncogenesis. Both of these proposals, that cancer can initiate either in a stem cell that has lost regulation of self-renewal or in a differentiated cell that has obtained the ability to self-renew, are supported by the identification of cancer stem cells. These are rare cells within a tumor that have the ability to self-renew and to give rise to phenotypically diverse cancer cells. Some evidence suggests that they drive tumorigenesis.

It has been shown in several types of cancer that tumors are maintained in a growing cancerous state by only a small fraction of particular tumor cells. These cells have surface proteins called markers, which are characteristic of the stem cell normally present in the tissue. For example, it has been established that only about one in a million acute myeloid leukemia cells can develop into new leukemias when transferred *in vivo* and that these cells expressed the same markers (CD34+; CD38−) as normal hematopoietic stem cells. Also, brain cancer stem cells display normal neural stem cells markers. Interestingly, the proportion of brain cancer stem cells identified in a variety of brain cancers correlates with the course of the disease, or prognosis. Fast-growing tumors such as glioblastomas had more brain cancer stem cells than slow-growing tumors like astrocytomas. Evidence for the existence of breast cancer stem cells was obtained by testing whether human breast cancer cells could give rise to new tumors when grown in immunocompromised mice (Al-Hajj, 2003). A minority subpopulation of breast cancer stem cells was isolated based on the expression of cell surface markers (CD44+

CD24$^{-/low}$) and these cells showed a 10- to 50-fold increase in ability to form tumors in animals, compared with the bulk of breast tumor cells. Furthermore, these cells were not only able to demonstrate the ability to self-renew but were also able to give rise to cells with different characteristics or phenotypes that made up the bulk of the tumor. These observations support the concept these cells are breast cancer stem cells.

Mammary stem cells react to physiological cues, such as hormones, to provide a source of proliferation and differentiation during pregnancy for the creation of a milk-generating breast. A major factor that protects women from breast cancer is an early, first, full-term pregnancy (Chapter 9). It is suggested that the depletion of stem cells as a result of the burst of differentiation that occurs during pregnancy is the reason why pregnancy is protective against breast cancer. There may be fewer breast stem cells that have the potential of becoming breast cancer stem cells over time in women who have had children in early adulthood.

In summary, a small minority of cancer stem cells may drive tumorigenesis in some cancers, similar to the small number of adult stem cells that drive the growth of normal tissues.

Molecular mechanisms of self-renewal

Let us examine the molecular mechanisms of self-renewal. The molecular mechanisms that regulate self-renewal of stem cells are just beginning to be understood. There is some evidence that the Wnt signaling pathway, which is important for regulating pattern formation during development, may be involved in the self-renewal process of stem cells during development, in the adult, and also in cancer. When the Wnt regulated transcription factor, Tcf (see below) is deleted by a gene knockout in mice, animals born lack stem cells of the intestines. In addition, hematopoietic stem cells respond to Wnt signaling *in vivo* and require Wnt signaling for self-renewal (Reya *et al.*, 2003). Data from DNA arrays show that the gene expression pattern in response to Wnt signaling is similar between colon stem cells and colon cancer cells, but differs in differentiated colon cells, suggesting that Wnt signaling plays a role in stem (cancer) cell self-renewal. The Hedgehog signaling pathway too, which is also important for regulating pattern formation in the embryo, has been implicated in the process of stem cell self-renewal as well. Both the Wnt and Hedgehog signaling pathways will be discussed below.

◎ The Wnt signaling pathway

Wnt proteins (>19 members) are secreted intercellular signaling molecules that act as a ligand to trigger a specific signal transduction

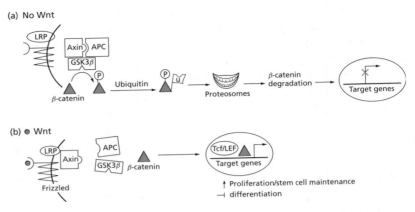

Figure 7.2 The Wnt signaling pathway

pathway (Figure 7.2). It is easiest to examine the cell in the absence of Wnt ligand first (Figure 7.2a). In this state, several proteins associate together in the cytoplasm to form a degradation complex. The degradation complex consists of glycogen synthase kinase (GSK3β), axin, and adenomatous polyposis coli (APC) protein. Axin and APC form a structural scaffold for GSK3β, which is a serine/threonine kinase. This complex modifies an important transcriptional coactivator called β-catenin via phosphorylation and ubiquitination. These modifications act as molecular flags that target β-catenin for degradation by the proteosome. Since β-catenin is not available in unstimulated cells, target genes under the regulation of β-catenin are repressed.

Upon binding of Wnt ligand to its seven-pass transmembrane receptor, Frizzled, and co-receptor LRP (low-density lipoprotein receptor related protein), axin is recruited to co-receptor LRP and this disrupts the assembly of the degradation complex (Figure 7.2b). In addition, an inhibitor of GSK3β, dishevelled protein, is activated via phosphorylation (not shown). These events allow β-catenin to escape degradation and move into the nucleus where it can act as a co-activator of the Tcf/LEF (T-cell factor/lymphoid enhancing factor) family of transcription factors to regulate specific target genes (e.g. c-myc, cyclin D, and adhesion molecules from the EPH-family). See Pause and Think.

Self test Close this book and try to redraw Figure 7.2. Check your answer. Correct your work. Close the book once more and try again.

Wnt signaling and cancer

*Wnt*1 was one of the first proto-oncogenes discovered. Viral integration-induced oncogene activation and subsequent cancer of the mammary gland. Mutations that constitutively activate the Wnt signaling pathway have been identified in several particular cancers. For example, such mutations are responsible for 90% of colorectal cancer. This translates in human terms to 50,000 lives in the USA alone per year. Most of the mutations inactivate the function of APC or activate β-catenin, but rarely alter the ligand Wnt. Colorectal cancer can be classified into two forms: familial forms and sporadic forms. Patients with the inherited cancer predisposition syndrome, familial adenomatous polyposis coli (FAP), carry a germline mutation in the APC gene and develop high numbers of polyps in the colon (polyposis) in early adulthood. As a result of having many polyps, these patients have an increased risk of colorectal cancer. The *APC* gene acts as a true tumor suppressor gene in that both copies of the *APC* gene are inactivated in colorectal tumors. Most mutations occur in the coding sequences for the central region of the APC protein (codons 1250–1500) referred to as the mutation cluster region, in both germline and somatic cases (Figure 7.3). See Pause and Think.

The intestines is a highly regenerative tissue whereby stem cells and epithelial progenitors that reside in the crypts give rise to more differentiated cells that migrate along the villi (Figure 7.4a; stem cells shown in red). Upon reaching the top of the villi, fully differentiated cells undergo apoptosis. Normally, Wnt signaling is required for maintaining the stem cell characteristics of the crypt cells. It has been proposed that acquiring constitutive activation of Wnt leads to colon cancer stem cells. Alternatively, one study reported that *APC* mutations associated with colorectal cancers were only present in dysplastic cells found at the top of the crypts in precancerous lesions (adenomas), and not in the stem cells at the base of the crypt (Shih *et al.*, 2001). Microscopic examination of precancerous lesions revealed an abrupt transition between the dysplastic compartment at the top of the crypts and the normal epithelium at the bottom of the crypts. The researchers propose that the morphogenesis of a colorectal tumor occurs in a top-down direction as illustrated in Figure 7.4b (dysplastic cells shown in red).

Mutations in the Wnt signaling cascade also promote other types of cancers. Activating mutations of β-catenin that affect the regulatory sequences essential for its targeted degradation can lead to skin tumors. Mutations

> **PAUSE AND THINK**
>
> What is the ultimate molecular consequence of having inactivating mutations in *APC*? Constitutive activation of Tcf transcriptional activators.

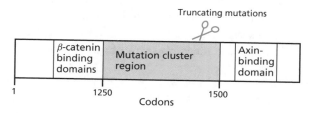

Figure 7.3 The mutation cluster region of APC

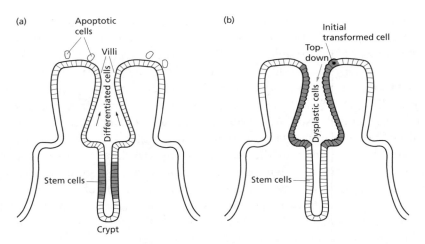

Figure 7.4 (a) Stem cells reside in the crypts of intestinal villi (b) Colorectal tumors may initiate at the top of the crypts

in the axin gene are found in hepatocellular carcinoma. Many of the axin gene mutations lead to protein truncations that delete the axin-β-catenin binding sites. Therefore, these observations suggest that some transforming mutations may function to reactivate the self-renewal pathway. The cells carrying these mutations can be thought of as *de novo* stem cells, that is cells that have acquired stem cell characteristics as a result from mutation, and were not produced from self-renewal of other stem cells.

The Hedgehog signaling pathway

The Hedgehog (Hh) signaling pathway also plays important roles in embryonic development, tissue self-renewal, and carcinogenesis. It is essential for pattern formation in many tissues including the neural tube, skin, and gut. Similar to Wnt proteins, Hh proteins (three members: Sonic, Desert, and Indian) are secreted intercellular signaling molecules that act as a ligand to trigger a specific signal transduction pathway (Figure 7.5). Two transmembrane proteins, Patched and Smoothened (related to Frizzled described above), are responsible for signal transduction by Hh. In the absence of Hh (Figure 7.5a), Patch inhibits Smoothened and thus suppresses the pathway. Upon binding of Hh to Patched (Figure 7.5b), inhibition of Smoothened is relieved. The signal is transduced into the cell and causes a large microtubule complex to dissociate and release the Gli zinc finger transcriptional activators so that they can be translocated to the nucleus to regulate the expression of target genes.

Self test   Close this book and try to redraw Figure 7.5. Check your answer. Correct your work. Close the book once more and try again.

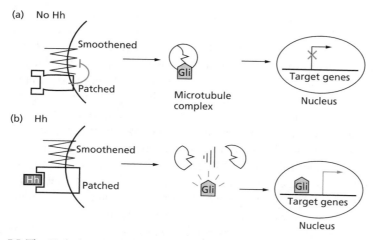

Figure 7.5 The Hedgehog signaling pathway

A leader in the field of differentiation: Cliff Tabin

Cliff Tabin once told me that he never really wanted a science career that required only working at the bench; he preferred to formulate hypotheses about molecular developmental biology and to teach. He is now renowned worldwide for his discovery of the protein Sonic Hedgehog, and the many functional studies of this protein that have followed.

Drosophila larvae that carry mutations in a specific gene have a phenotype characterized by bristles. Hence, the name given to the protein product of that gene was hedgehog. The vertebrate homolog was named Sonic Hedgehog by the Tabin laboratory after the character from a British comic book. Coincidently their scientific paper describing the cloning of the vertebrate gene came out in the same year as the US release of the Sega Sonic Hedgehog video game. Scientists do have fun at play.

Cliff carried out his Ph.D. in Robert Weinberg's laboratory at M.I.T, Boston. His postdoctoral tenure was carried out at the Department of Biochemistry at Harvard University and the Department of Molecular Biology at Massachusetts General Hospital. Cliff is currently a Professor of Genetics at the Harvard Medical School in Boston.

Hedgehog signaling and cancer

Patched is defined as a tumor suppressor gene; patients with Gorlin's syndrome carry a germline mutation in one copy of Patched and have a predisposition to develop skin, cerebellar, and muscle tumors (basal cell carcinoma (BCC), medulloblastomas, and rhabdomyosarcomas respectively). Inactivating mutations in Patched and activating mutations of Smoothened were identified in sporadic (opposed to familial) human BCC tumors and Gli-1 expression was found in nearly all BCCs. In fact, all sporadic BCCs possess an activated Hh signaling pathway.

It has been suggested that since BCC are tumors that include hair-follicle differentiation, they may originate from a source of stem cells that reside in a structure of the hair follicle, called the hair follicle bulge, which has acquired an aberrant Hh signaling pathway. These cancer stem cells could self-renew and give rise to the differentiated hair follicle cells observed in BCC tumors. Similarly, it has been proposed that medulloblastoma, the most common childhood malignant brain tumor, arises from interneuron precursors that possess an inappropriately activated Hh pathway. Activation of this pathway by mutation is observed in 30% of sporadic medulloblastomas. Molecular evidence of an activated Hh pathway was also reported for gliomas. Gli was originally identified as an amplified gene in cultured glioma cells. Unlike Wnt, where mutations involved in carcinogenesis rarely affect this ligand, Hh is overexpressed in upper gastrointestinal tumors. Taken together, it is apparent that the Hh signal pathway is relevant for a several types of human cancer.

Additional properties of stem cells and tumor cells

Cancer stem cells, in addition to being able to self-renew, also have the ability to give rise to more differentiated cell types with limited proliferative capacity. Teratocarcinoma is an obvious example whereby the tumor contains undifferentiated stem cells and nonproliferative differentiated cells such as bone and cartilage. Although the degree of differentiation for other cancers is less obvious, cancer stem cells from other cancers have been shown to generate more cells with different phenotypes. The cell heterogeneity present in a tumor may be derived from cancer stem cells that not only can self-renew but can simultaneously give rise to more differentiated cells. A striking demonstration has suggested that tumor cells can be reprogrammed to become normal and totipotent during experimental manipulation (like fully differentiated cells during cloning experiments). When placed in early mouse embryos, teratocarcinoma cells can mimic stem cells and can contribute to normal development. In addition, the degree of differentiation of a cell may also affect the outcome of oncogene activation. The use of selective gene promoters to drive the expression of a Ras oncogene in different cell populations with different degrees of differentiation (e.g. stem cells or committed progeny cells) resulted in tumors with different malignant potential (i.e. malignant carcinomas vs. benign papillomas). Although additional studies are needed, this suggests that the activation of an oncogene may be carcinogenic in some states of differentiation and not in others within a particular cell lineage.

Another feature of stem cells that is shared with tumor cells is the ability to migrate to other tissues of the body. In contrast, most differentiated cells remain localized to a specific tissue. Transplantation of stem cells has demonstrated the migratory nature of stem cells. For example, bone marrow cells migrate to several nonhematopoietic tissues such as the brain, liver and lung. Metastasis of tumor cells from a primary site to secondary sites is a characteristic that makes cancer lethal (details are discussed in Chapter 9). It has been proposed that the origin of the transformed cell determines the potential for metastasis: tumors arising from a stem cell are more likely to metastasize compared with tumors arising from more differentiated cells which are less likely to spread. The inherent ability of a stem cell to migrate may cause these cells to be aggressively metastatic, if transformed. The myeloid leukemias support this view: transformed stem cells are likely to be malignant while transformed committed progenitor cells are likely to be benign.

Moreover, telomerase activity (discussed in Chapter 3), which is necessary for tumor proliferation and progression and is present in 90% of human cancers, is present in normal stem cells and proliferative cells. This supports the hypothesis that cancer cells are derived from normal stem cells.

Summary

This discussion leads to the question of the nature of the cell that initiates carcinogenesis. It may be hypothesized that tumors arise from stem cells within a tissue or alternatively from more differentiated cells that acquire the stem cell quality of self-renewal. There is supporting evidence for both. Alternatively, a continuum of target cells relative to different states of differentiation may exist: stem cells, progenitor cells, and terminally differentiated cells may all be targets for transformation. Further, the stage of differentiation of the target cell may affect the malignant potential and severity of the cancer.

7.2 Role of lineage-specific transcription factors in blocked differentiation

Acute myeloid leukemia serves as an important paradigm for examining how disruption of transcription-factor function can interfere with differentiation, and lead to cancer.

Acute myeloid leukemia (AML) is a disease characterized by a block in the differentiation of the granulocyte or monocyte lineage (Figure 7.6). There are several subtypes of acute myeloid leukemia. The classification system of this disease is still evolving but should eventually reflect

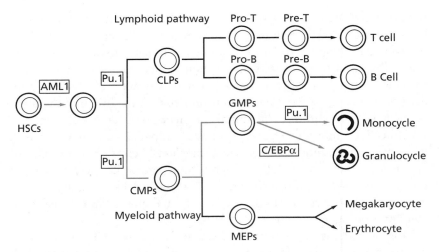

Figure 7.6 Haematopoietic lineages: disruption in the granulocyte or monocyte lineage (shown in red) leads to AML. From *Nature Rev. Cancer*, Vol. 3, Tenen, D.G., "Disruption of differentiation in human cancer: AML shows the way", pp. 89–101. Copyright 2003, with permission from D.G. Tenen

molecular features at the point of the differentiation block. The lineage is organized as a hierarchy that begins with pluripotent hematopoietic stem cells (HSCs). These cells self-renew and form progenitor cells. The progenitor cells differentiate into several types of precursor cells including myeloid precursor cells. Myeloid precursor cells are common to both the monocytes and granulocyte lineages. Several transcription factors have been identified to be important in hematopoietic lineage development. One factor, AML1, is involved in almost all lineages. Others are lineage-specific factors (differentiation factors), such as PU.1 and CCAAT/enhancer binding protein α (C/EBPα). Lineage specific transcription factors activate a particular set of lineage-specific genes and/or inhibit the cell cycle for terminally differentiated cells. PU.1 is involved in the differentiation of the common myeloid progenitor cell (CMP) and, later on, in the differentiation of monocytes/macrophages. Most myeloid specific genes have PU.1 sites in their promoters. C/EBPα, a zinc finger transcription factor, functions in the differentiation of granulocytes.

Many mutations that are typically found in acute myeloid leukemia affect specific transcription factors; both chromosomal translocations [e.g. t(8;21)] and coding region mutations are common. The gene for the AML1 transcription factor is disrupted in the t(8;21) translocation and this translocation leads to acute myeloid leukemia. The chromosomal translocation, t(8;21) is identified in both hematopoietic stem cells and more differentiated cells in patients, thus providing additional evidence that the transforming mutations of acute myeloid leukemia occur in hematopoietic stem cells.

Mutations in lineage-specific transcription factors are found in patients with acute myeloid leukemia subtypes that are consistent with their role in normal hematopoiesis. PU.1 mutations are found in the earliest stage (M0: very immature leukemia) and in monocytic leukemias reflecting PU.1's early role in myeloid precursor cells and in the development of monocytes/macrophages. Approximately 10% of acute myeloid leukemia patients carry a mutation in c/EBPα and most of these cases are associated with the granulocytic subtype reflecting c/EBPα's role in granulocyte differentiation. Thus, mutation of transcription factors involved in differentiation is an important mechanism behind oncogenesis.

In acute myeloid leukemia, most leukemic cells have a limited capacity for proliferation but are replenished by rare leukemic stem cells (Section 7.1). Therefore, the self-renewal ability of stem cells is important in the maintenance of this disease. A transcriptional repressor, Bmi-1, has been demonstrated to be essential for the control of self-renewal in hematopoietic stem cells and in leukemic stem cells (Lessard and Sauvageau, 2003). *In vitro*, leukemic stem cells that lack Bmi-1 show growth arrest in the G1 phase of the cell cycle and begin to differentiate. *In vivo*, mice with a Bmi-1 gene knockout show a progressive depletion of all blood cells indicating Bmi-1's essential role in hematopoietic stem cells. In addition, in mouse models using leukemic stem cells lacking Bmi-1, lower numbers of leukemic cells are detected in the peripheral blood compared with controls, indicating that these cancer stem cells have proliferative defects. This is an example of how a common gene can regulate self-renewal in both normal and cancer stem cells. Bmi-1 normally exerts its effects partially by repressing the expression of two cyclin-dependent kinase inhibitors p16 and p14 via chromatin remodeling. Bmi-1's role as a human oncogene is supported by the identification of Bmi-1 gene amplification in some mantle cell lymphomas.

Acute promyelocytic leukemia, a subtype of acute myeloid leukemia, is most often characterized by the chromosomal translocation, t(15;17), that results in the fusion of the PML gene with the retinoic acid receptor alpha (RARα) gene to create a hybrid protein with altered functions. As described in Chapter 3, RARs (α, β, and γ) are members of the steroid hormone receptor superfamily and act as ligand dependent transcription factors that are important effectors of retinoic acid's essential role in cell differentiation. The wild-type receptors bind to the retinoic acid response element in target genes as RAR-RXR heterodimers. In the absence of retinoic acid (RA), the receptors associate with HDAC-co-repressor complexes that silence target genes by histone deacetylation and subsequent chromatin compaction (Figure 7.7a). Upon binding of RA, the receptor undergoes a change in shape that causes the receptor to dissociate from the HDAC-co-repressor complex and allows the receptor to interact with co-activators in order to transcriptionally

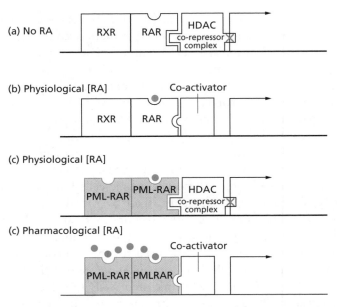

Figure 7.7 The activity of RAR and the PML-RAR fusion protein vs. retinoic acid concentration

induce its target genes (Figure 7.7b). Co-activators recruit HATs and also mediate interactions with the basal transcriptional machinery. The oncogenic fusion protein (shown in red) maintains both the DNA binding domain and the ligand binding domain of the RARα receptor. It has a higher affinity for HDAC and does not dissociate in the presence of physiological concentrations of RA (Figure 7.7c). In addition, it is likely that the fusion protein forms homodimers that may act in a dominant negative manner by blocking wild-type RAR-RXR heterodimers or by recruiting novel co-repressors. The normal role of the PML protein may also be disrupted in the fusion protein. PML protein is normally found in nuclear organelles called nuclear bodies and, as a co-activator of p53, acts as a pro-apoptotic protein. Thus, although altered gene expression of retinoic acid target genes is most likely the predominant mechanism of PML-RAR's oncogenic ability, additional possibilities such as affecting PML function, exist (Salomoni and Pandolfi, 2002).

◎ Therapeutic strategies

The concept of cancer stem cells has important implications for the design and testing of new cancer drugs. First, since cancer stem cells support the growth and migration of the tumor, drugs need to target this small subset of cells within the tumor. Many existing drugs give hopeful initial responses that are followed by disappointing latter reoccurrences. Drugs targeted at cancer stem cells may prevent reoccurrence and actually cure metastatic cancer.

PAUSE AND THINK

Consider the possible side-effects of drugs that target cancer stem cells? Such drugs may destroy normal stem cells. Depending on the tissue, this may or may not be acceptable. For example, the loss of breast stem cells may be acceptable for most patients since breast cancer usually strikes after child-bearing years and the breast is not a vital organ. On the other hand, destruction of skin stem cells would result in serious problems since the skin would be unable to self-renew.

In addition, the efficacy of such new drugs should be determined by its effect on the cancer stem cell population and not on overall tumor regression. Drugs may be successful at killing all of the cells of a tumor except cancer stem cells, so that measuring tumor regression would not reflect the fact that the most dangerous tumor cell types remained unaffected. Differences in drug resistance between cancer stem cells versus other tumor cells is a possible explanation for such a scenario. Below is a sample of drug strategies that target self-renewal or differentiation pathways.

7.3 Inhibitors of the Wnt pathway

The importance of the Wnt pathway in several cancers, particularly colorectal cancer, suggests that the molecular components of this pathway are good targets for new therapeutics. See Pause and Think.

Disruption of the protein–protein interaction between β-catenin and the Tcf transcription factors (Figure 7.8) is one strategy that has been investigated (Lepourcelet *et al.*, 2004). This interaction occurs downstream of the APC degradation complex and is the end-point effect of Wnt signaling. Drugs acting at this stage would counter both inactivating mutations in APC, Axin, GSK3β, and activating mutations in β-catenin that cause inappropriate formation of β-catenin/Tcf complexes. The observation that TCf4 inhibition induces the differentiation of colorectal cancer cells into epithelial villi is evidence that supports this approach. Lepourcelet *et al.* (2004) identified three natural compounds from a high-throughput screen that acted as inhibitors of the β-catenin-Tcf interaction. Also, these compounds that share a core chemical structure inhibited the expression of two Tcf-target genes and inhibited colorectal cancer cell proliferation. Although in its early stages, this

---

PAUSE AND THINK

What molecules would you target in this pathway?

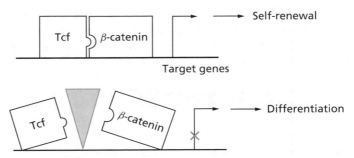

Figure 7.8 Drug strategy to inhibit the β-catenin–Tcf interaction

strategy holds promise for the development of new cancer therapeutics. Since these drugs target a molecular pathway that is important in self-renewal, they have a greater chance of tumor eradication rather than only tumor regression.

7.4 Inhibitors of the Hh pathway

Inhibitors of the Hh signaling pathway are being investigated as cancer therapeutics (Figure 7.9). One example is cyclopamine, a steroidal alkaloid that is found in high levels in wild corn lilies. [Its name comes from the cyclopic effects (formation of a single eye) observed by its action as a teratogen. Pregnant sheep that ingested high quantities of wild corn lilies gave birth to cyclopic lambs. There are obvious implications for treating woman of child-bearing age with this teratogen.] Cyclopamine suppresses the Hh pathway by inhibiting the activity of the transmembrane protein, Smoothened. In animal studies, cyclopamine treatment blocked growth of medulloblastomas cells and affected the regulation of genes involved in differentiation (Berman *et al.*, 2002). Expression of a neuronal stem cell marker, neurofilament nestin, was decreased while expression of a marker of neuronal differentiation, Neuro D, was increased.

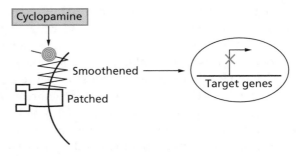

Figure 7.9 Inhibition of the Hedgehog pathway by cyclopamine

Genetech is involved in the development of small molecule antagonists and Hedgehog blocking antibodies as means of inhibiting the Hh pathway for new cancer treatments.

7.5 Leukemia and differentiation therapies

Differentiation therapy aims to promote the maturation and differentiation of cells such that a malignant phenotype changes into a benign phenotype. The mechanisms may include the induction of growth arrest and apoptosis via gene regulation. The use of differentiation therapy for the treatment of acute promyelocytic leukemia (APL), a subtype of AML, has been one of the great success stories of the last few decades. Retinoid treatment, using all-trans retinoic acid (ATRA) has transformed a deadly leukemia into one of the most treatable forms of cancer. Retinoid therapy along with chemotherapy results in complete remission with 80% survival at 5 years. Note that treatment with retinoic acid alone causes retinoic acid syndrome in 10–15% of patients but administration with chemotherapy reduces this side effect. The high concentration of retinoids administered "push" RA binding to the ligand binding domain of the RARα-PML fusion protein (Figure 7.7d). This results in a conformational change that induces the exchange of the HDAC-co-repressor complex for a co-activating complex. Consequently, the target genes of RARα are expressed and the block of differentiation is overcome. The gene encoding the differentiation specific transcription factor C/EBP is one of the targets of retinoid treatment.

■ **CHAPTER HIGHLIGHTS—REFRESH YOUR MEMORY**

- Stem cells provide a source of cells for differentiation.

- Stem cells are characterized by their ability to self-renew and to form more differentiated progeny, simultaneously.

- Self-renewal is a characteristic that is shared with tumor cells.

- Stem cells are more likely to accumulate mutations compared with other cells.

- Cancer stem cells are rare cells within tumors that self-renew and drive tumorigenesis.

- Evidence suggests that the Wnt signaling pathway is involved in self-renewal.

- The transcriptional co-activator, β-catenin, is stabilized in the presence of Wnt.

- Mutations that inappropriately activate the Wnt signaling pathway promote carcinogenesis, particularly colon cancer.

- A germline mutation in the APC gene causes familial adenomatous polyposis.

- The Hedgehog signaling pathway is also implicated in self-renewal.

- Hh signaling exerts its effects via the Gli zinc finger transcription factors.

- Inappropriate activation of the Hh pathway is linked to many cancers.

- A germline mutation in the Patched gene causes Gorlin's syndrome.

- Both stem cells and malignant tumor cells migrate to other tissues in the body.

- The degree of differentiation of the transformed founder cell may determine metastatic potential.

- Cancer may originate from stem cells or may involve reactivation of the self-renewal process.

- AML is an important paradigm for the role of differentiation in cancer.

- Lineage-specific transcription factors are commonly mutated in AML.

- Drugs that target the self-renewal pathways of cancer stem cells are more likely to achieve a cure.

- Interference with both the Wnt and Hh signaling pathways are recent therapeutic strategies being explored.

- Retinoid therapy is successful as a differentiation therapy for APL.

ACTIVITY

The tumor suppressor gene Pten has recently been found to be an important regulator of proliferation of neural stem cells. Critically analyze the evidence that supports this and discuss its relevance to specific cancers. [Hint: start with *Nature Med.* **8**:16 (2002).]

FURTHER READING

Altucci, L. and Gronemeyer, H. (2001) The promise of retinoids to fight against cancer. *Nature Rev. Cancer* 1:181–193.

Giles, R.H., van Es, J.H. and Clevers, H. (2003) Caught up in a Wnt storm: Wnt signaling in cancer. *Biochem. Biophys. Acta* **1653**:1–24.

Owens, D.M. and Watt, F.M. (2003) Contribution of stem cells and differentiated cells to epidermal tumors. *Nature Rev. Cancer* 3:444–451.

Pardal, R., Clarke, M.F. and Morrison, S.J. (2003) Applying the principles of stem-cell biology to cancer. *Nature Rev. Cancer* 3:895–902.

Perez-Losada, J. and Balmain, A. (2002) Stem-cell hierarchy in skin cancer. *Nature Rev. Cancer* 3:434–443.

Polakis, P. (2000) Wnt signaling and cancer. *Genes Dev.* **11**:1837–1851.

Reya, T., Morrison, S.J., Clarke, M.F. and Weissman, I.L. (2001) Stem cells, cancer, and cancer stem cells. *Nature* **414**:105–111.

Ruiz I Altaba, A., Sanchez, P. and Dahmane, N. (2002) Gli and Hedgehog in cancer: tumours, embryos and stem cells. *Nature Rev. Cancer* 2:361–372.

Smalley, M. and Ashworth, A. (2003) Stem cells and breast cancer: a field in transit. *Nature Rev. Cancer* 3:832–844.

Tenen, D.G. (2003) Disruption of differentiation in human cancer: AML shows the way. *Nature Rev. Cancer* 3:89–101.

■ WEB SITES

Wnt Home Page www.standford.edu/~nusse/wntwindow.html

■ SELECTED SPECIAL TOPICS

Al-Hajj, M., Wicha, M.S., Benito-Hernandez, A., Morrison, S.J. and Clarke, M.F. (2003) Prospective identification of tumorigenic breast cancer cells. *PNAS* **100**:3983–3988.

Berman, D.M., Karhadkar, S.S., Hallahan, A.R., Pritchard, J.I., Eberhart, C.G., Watkins, D.N., Chen, J.K., Cooper, M.K., Taipale, J., Olson, J.M. and Beachy, P.A. (2002) Medulloblastoma growth inhibition by hedgehog pathway blockade. *Science* **297**:1559–1561.

Lepourcelet, M., Chen, Y.-N.P., France, D.S., Wang, H., Crews, P., Petersen, F., Bruseo, C., Wood, A.W. and Shivdasani, R.A. (2004) Small-molecule antagonists of the oncogenic Tcf/beta-catenin protein complex. *Cancer Cell* **5**:91–102.

Lessard, J. and Sauvageau, G. (2003) Bmi-1 determines the proliferative capacity of normal and leukemic stem cells. *Nature* **423**:255–260.

Reya, T., Duncan, A.W., Ailles, L., Domen, J., Scherer, D.C., Willert, K., Hintz, L., Nusse, R. and Weissman, I.L. (2003) A role for Wnt signaling in self-renewal of haematopoietic stem cells. *Nature* **423**:409–414.

Salomoni, P. and Pandolfi, P.P. (2002) The role of PML in Tumor Suppression. *Cell* **108**:165–170.

Shih, I.-M., Wang, T.-L., Traverso, G., Romans, K., Hamilton, S.R., Ben-Sasson, S., Kinzler, K.W. and Vogelstein, B. (2001) Top-down morphogenesis of colorectal tumors. *PNAS* **98**:2640–2645.

8 Metastasis

Introduction

Most cells of the body normally remain resident within a particular tissue or organ. Liver cells remain in the liver and cannot be found in the lung and vice versa. Organs have well-demarcated boundaries defined by surrounding basement membranes. Basement membranes are acellular structures made up of a fabric of extracellular matrix (ECM) proteins: predominantly laminins, type IV collagen, and proteoglycans. Cancer is distinctly characterized by the spreading of tumor cells throughout the body. The process by which tumor cells migrate from a primary site to other parts of the body is called metastasis. Metastasis is the fundamental difference between a benign and malignant growth and represents the major clinical problem of cancer. A primary tumor can be surgically removed relatively easily, whereas once hundreds or more metastases have been established throughout the body they are practically impossible to remove. Sadly, over 50% of solid tumors have metastasized at the time of diagnosis.

PAUSE AND THINK

Why is the spread of cells throughout the body lethal?

The spread of cells throughout the body results in physical obstruction, competition with normal cells for nutrients and oxygen, and invasion and interference with organ function. Interestingly, specific cancers metastasize to particular sites. Many of the preferences observed for the spread of specific cancers to specific metastatic locations can be explained by the directionality of blood flow. Since the blood stream is the predominant means of long-distance transport, organs in close proximity "en route" are likely to be main sites of metastasis for a particular primary tumor. However, about one third of the location of frequent metastases is puzzling in this regard. For example, breast cancer metastasizes to bone more frequently than anatomy would suggest. One explanation of this observation was described over 100 years ago in the "seed and soil" theory proposed by Paget. It described cancer cells as "seeds" requiring a match with optimal environments or "soils" to succeed. Recent molecular observations suggest that receptors lining the capillaries in the organs to which cancer spreads influences the destination of metastasized cells, and these findings support the "seed and soil" theory. The ability of cancer cells to metastasize is dependent on the interactions of their cell surface molecules with the microenvironment, including neighboring cells and the extracellular matrix. Metastasizing cells break free from their neighboring cells and cross the physical boundaries they encounter. Although cancers are largely successful in metastasizing in the long run, on the cellular level, only 1 in 10,000 metastazing cells survive transport.

8.1 Steps of metastasis

There are several major steps involved in metastasis: migration, intravasation, transport, extravasation, and metastatic colonization (Figure 8.1). The first step (migration) and the last step (metastatic colonization) have been demonstrated to be rate limiting and their rate dictates the overall metastatic ability of the cancer. As we examine each of the major steps below, try to evaluate the molecular components as possible therapeutic targets and think about strategies that may be used to develop new drugs. Examples of therapeutic strategies will be described at the end of the chapter.

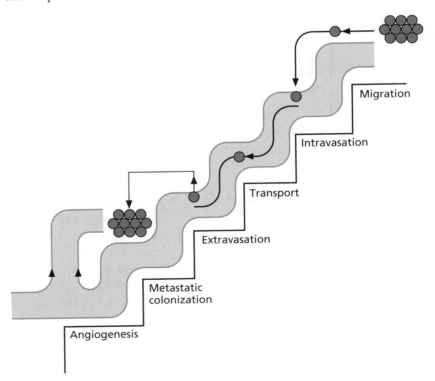

Figure 8.1 Steps of metastasis

8.2 Tools of cell migration: cell adhesion molecules, integrins, and proteases

Cell adhesion molecules

In order for cells to migrate away from the primary tumor (top right, Figure 8.1), cells must break free from the normal molecular constraints that link adjacent cells to each other. Cell adhesion molecules (CAMs) and cadherins are two families of proteins that mediate homotypic (same cell type) and heterotypic (different cell types) recognition. They "hook" cells into place extracellularly (Figure 8.2). Cadherins (Figure 8.2a) are calcium dependent transmembrane glycoproteins that interact, via catenins (Figure 8.2b), with the cytoskeleton (Figure 8.2c).

Catenins also bind to transcription factors and induce gene expression in the nucleus. Thus intercellular interactions are networked to mediators of intracellular functions. Several lines of evidence suggest that these molecules are important during metastasis. Cells treated with antibodies that block the function of cadherins became invasive in collagen gels, indicating an increased metastatic potential. Furthermore, transfection of

> **PAUSE AND THINK**
>
> Do you think the cytoskeleton is important for cell migration? Why?

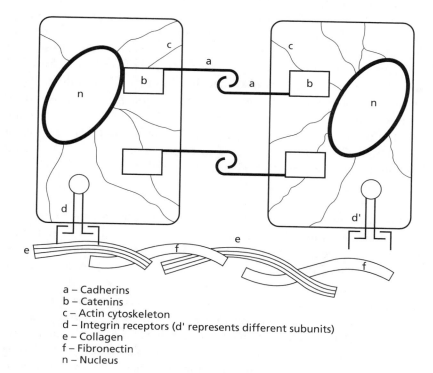

a – Cadherins
b – Catenins
c – Actin cytoskeleton
d – Integrin receptors (d' represents different subunits)
e – Collagen
f – Fibronectin
n – Nucleus

Figure 8.2 Cell adhesion molecules and associated components

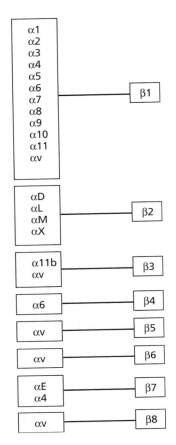

Figure 8.3 The integrin family: α and β subunit heterodimers

the *E-cadherin* gene into metastatic epithelial cells can render them non-invasive. Mutations in the extracellular domain and methylation in the promoter region of the *E-cadherin* gene, the gene encoding the predominant cell adhesion molecule in epithelial cells, have been identified in gastric and prostate carcinomas. It has been suggested that E-cadherin acts as a tumor suppressor that normally functions to secure cell–cell adhesion and suppresses metastasis of tumor cells to distant sites.

Integrins

Cells must also break free from the normal molecular constraints with the ECM. Integrin receptors (Figure 8.2d) are a family of heterodimers (>24) made up of a range of α and β subunits (Figure 8.3) that mediate cell–ECM interactions and intracellular signal transduction. The recognition of the different components of the ECM (e.g. collagen (Figure 8.2e), fibronectin (Figure 8.2f), or laminin) depends on the α and β subunit composition. Many ligands for integrin receptors contain an Arg-Gly-Asp sequence that is involved in binding. Upon ligand binding, the integrins cluster in the membrane and affect the cytoskeleton through interaction with actin binding proteins and specific kinases, such as focal adhesion kinase (FAK). In contrast to most transmembrane receptors, the cytoplasmic tail of integrins do not exhibit any catalytic activity (e.g. kinase activity) themselves. Data suggest that FAK mediates cell motility through recruitment of Src and activation of the RAS pathway (discussed in Chapter 4). In addition to this typical outside of the cell to inside of the cell signaling, integrins also mediate "inside-out" signaling. Intracellular signals mediated at the cytoplasmic domain of integrins induce a conformational change in the extracellular domain and thus regulate the affinity of the integrins for their ECM ligands. Integrins also have a role in anoikis, apoptosis triggered in response to lack of ECM ligand binding. Integrins without suitable ECM ligands recruit caspase 8 to the membrane and trigger apoptosis. Thus, integrin dependent cell anchorage is crucial for survival of the cell.

Altered integrin receptor expression in tumor cells can enable the mobility of metastazing cells by modifying membrane distribution and/or allowing adherence to different ECMs. Regulation by the cancer cell must result in precise intermediate strengths of adhesiveness to produce the maximum rate of cell migration, allowing cells to advance their leading edge and to release their lagging edge. The role of integrins in motility is obvious in melonoma cells in which its invasive front edge shows a strong pattern of expression of integrin αvβ3 that is absent in preneoplastic melanomas. At the moment, the complexity of ligand binding affinity and variable ligand concentration available to the various receptors makes it difficult to predict the outcome of altered integrin expression with respect to migration; sometimes integrins are downregulated and in some contexts they are induced. An altered expression of a specific integrin heterodimer may be permissive for invasion. For example, an increase of α6β4, a laminin binding integrin, promotes invasion through the basement membrane and the laminin matrix often secreted by epithelial tumors. Further still, altered integrin expression may facilitate invading cells to overcome anoikis.

Proteases

Invasion of tumor cells into the surrounding tissue requires the action of specific proteases that degrade a path through the ECM and stroma. Serine proteases and matrix metalloproteases (MMPs) are two families that are important. Although some tumor cells can synthesize MMPs, more often tumor cells induce surrounding stromal cells to produce MMPs. One appropriately named protein called extracellular matrix metalloprotease inducer (EMMPRIN) is upregulated on the membrane of tumor cells and induces MMP production in adjacent stromal cells. The family of MMPs can not only degrade all structural components of the ECM, but also other proteins residing on the outside of cells (e.g. endothelial cell growth factors), and thus are likely to play an important role in metastasis, including angiogenesis (Sections 8.6 and 8.7 below). Normally, these zinc-dependent proteinases are tightly regulated at several levels in addition to gene expression. First, they are synthesized as latent enzymes and require proteolytic cleavage to be activated. Also, endogenous tissue inhibitors (TIMPs) regulate their function. A tip in the balance between MMPs and TIMPs can signal invasion. MMPs are upregulated in almost all tumors and their expression profile can indicate the degree of tumor progression in some cancers.

8.3 Intravasation

Intravasation is the penetration of a cell into a blood or lymphatic vessel. The process requires several steps: the tumor cell must attach to the stromal face of the vessel, degrade the basement membrane (absent in lymphatic vessels), and pass between the endothelial cells (transendothelial migration) into the bloodstream. The chick egg provides a model system for the study of intravasation (Figure 8.4). Malignant human cells can be inoculated through a window made in the shell of an egg onto the chorionic epithelium beneath which lies a rich bed of blood vessels (Figure 8.4, top red arrow). Successful intravasation of the human tumor cells into the bloodstream of the egg is followed by metastasis to distant

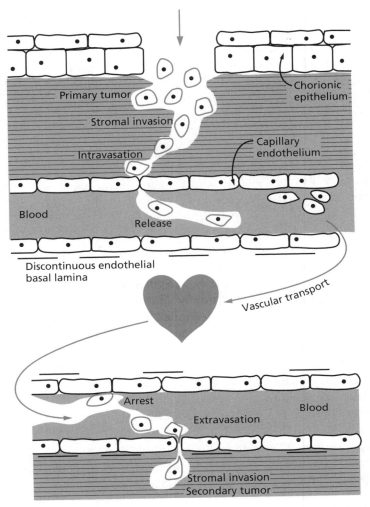

Figure 8.4 Model system for studying intravasation. Reprinted from *Cell*, Vol. 94, Quigley and Armstrong, "Tumor cell intravasation: *Alu*-cidated: the chick embryo opens the window", pp. 281–284, Copyright (1998), with permission from Elsevier

sites and subsequent formation of secondary tumors. Human cells from these secondary tumors can be detected by DNA amplification (Polymerase Chain Reaction; PCR) in samples taken opposite the inoculation site. Repetitive DNA sequences (approximately 300 bp long) called Alu sequences are present in the human genome and are not present in the chick genome. Genetically modified human cells created to test the role of a specific molecule in intravasation can be easily tested in this system. For example, cells experimentally altered to express reduced quantities of a particular protease receptor (urokinase receptor) results in reduced levels of Alu PCR product at sites opposite the inoculation site. This implies that this protease receptor normally plays a role in intravasation.

8.4 Transport

Transport through the bloodstream is one-way. Tumors cells travel singly or as clumps with platelets, called emboli, in the direction of blood flow. Specific cancers have favored sites of metastasis and this is partly due to the concept of the first pass organ. The first pass organ is the first organ en route via the bloodstream that lies downstream from the primary tumor site. The lung is the first pass organ for cells of the breast via the superior vena cava that also receives drainage from the lymphatic vessels. Thus, the lung is a common site of metastasis from breast cancer. The liver is the first pass organ for cells via the hepatic portal vein and is particularly vulnerable due to sinusoids, areas where blood is in direct contact with hepatocytes.

8.5 Extravasation

Extravasation is the escape of a tumor cell from a blood or lymphatic vessel. The steps involved are the same for intravasation but in reverse: the tumor cell must attach to the endothelium side of the blood vessel, pass through the endothelial cells and basement membrane, and migrate into the surrounding stroma.

Members of the selectin family of adhesion molecules, particularly E-selectin, are specifically expressed on endothelial cells and are important for the attachment of cancer cells to the endothelium. They are calcium-dependent transmembrane receptors that mediate interactions with cancer cells by binding to various glycoprotein ligands presented on adhering cells. Endothelial selectins are differentially expressed on the vasculature of different organs and may support the "seed and soil" theory discussed above. E-selectin expression in liver sinusoidal cells is triggered by Lewis lung carcinoma cells and may explain the preference of these cells to metastasize to the liver. Signaling between the selectins and their ligands appears to be bi-directional in that signal transduction in both participating cells has been demonstrated. That is, signaling initiates from both the selectin cytoplasmic tails and from their activated ligands. For example, cross-linking of E-selectin induces tyrosine phosphorylation in *endothelial cells* and also modifies endothelial cell shape. In contrast, stress-activated protein kinase-2 (SAPK2/p38), an isoform of MAPK, is induced in *cancer cells* upon binding of E-selectin (on endothelial cells) and is necessary for transendothelial migration (Laferriere *et al.*, 2002). This suggests that binding to E-selectin on the endothelium by cancer cells not only mediates adhesion to the endothelium but also triggers a signal transduction cascade that is important for transendothelial migration by the cancer cells.

8.6 Metastatic colonization

The words in the term "metastatic colonization" have been precisely chosen to describe the last stage of metastasis. Let us examine the "familiar" concept of colonization. The British established distant colonies in the New World, the growth of which were dependent on the surrounding waterways and harbors. As the settlers moved west, some of the new environmental conditions were unfavorable for growth. Other locations encouraged the development of a new means of water access and resulted in a flourishing settlement. Metastatic colonization is the establishment of a progressively growing tumor at a distant site, involving angiogenesis as an essential process to provide nutrients and oxygen. It is important to contrast this process to metastasized tumor cells that do not expand and remain dormant for years as micrometastases. Micrometastases maintain an overall balance between proliferation and apoptosis, and due to the lack of angiogenesis, do not demonstrate progressive growth. Metastatic colonization identifies the last step of metastasis that can be targeted to halt the complete clinical cancer phenotype.

A new class of genes, called metastasis-suppressor genes, have been identified by their low expression in metastatic cells compared with nonmetastatic tumor cells. Thus loss of function (analogous to the mechanism of tumor suppressor genes), increases the metastatic propensity of a cancer cell. Since the discovery of the first, *NM23*, seven additional genes have been identified to date, of which the protein products of some affect metastatic colonization. *MKK4* is one example. It is hypothesized that MKK4 protein induces apoptosis in response to the stress of a new microenvironment and thus suppresses metastatic colonization. The mechanism of action of metastasis suppressor proteins includes regulation of common signal-transduction pathways (e.g. MAPK) and gap–junction communication, though these are only beginning to be elucidated.

As alluded to above, metastatic colonization cannot be successful without the formation of new blood vessels. Angiogenesis is the process of forming new blood vessels from pre-existing blood vessels by the growth and migration of endothelial cells in a process called "sprouting". Although this process is common during embryogenesis, it rarely occurs in the adult, being reserved for wound healing and the female reproductive cycle. With respect to cancer, angiogenesis is essential for metastasized tumors since all cells must be within 100–200 μm of blood vessels (the diffusion limit of oxygen) in order to receive essential oxygen and nutrients. Sprouting of pre-existing vessels requires major reorganization involving destabilization of the mature vessel, proliferation and migration of endothelial cells, and maturation. It is regulated by the interaction of soluble mediators and their cognate receptors. Malignant cells in culture and host stromal cells induced by a tumor *in vivo* have been shown to be sources of these soluble mediators. The neovasculature that is formed in cancer is unlike that formed in wound healing. It is leaky and tortuous and provides direct entry, allowing cells easy access to the circulation. The neovasculature is also different at the molecular level from resting endothelium. For example, the integrins αvβ3 and αvβ5 are upregulated in angiogenic vessels compared with mature vessels. The proliferating endothelial cells of the sprouting vessel need to interact with components of the ECM that it is invading. The proangiogenic factors VEGF and bFGF induce the expression of these integrins. Molecular differences are not limited to the endothelium since the supporting pericytes and ECM show specific angiogenic markers (e.g. NG2 and oncofetal fibronectin respectively). Therefore, all components of angiogenic vasculature are molecularly distinct from normal vessels.

A leader in the field of angiogenesis: Judah Folkman

Judah's pioneering discoveries into the mechanism of angiogenesis opened up a new field of cancer research and supported his ground-breaking idea that tumors are dependent on angiogenesis. His laboratory identified the first angiogenic inhibitor and he is presently carrying out clinical trials of antiangiogenic therapies. He is also investigating the observation that some tumors remain dormant, sometimes indefinitely, due to the production by the tumor of an angiogenic inhibitor, but can become angiogenic when production of the inhibitor decreases.

Judah received his B.A. from Ohio State University and his M.D from Harvard Medical School. He has made advancements in a range of fields, including the development of the first atrioventricular implantable pacemaker and implantable polymers for controlled release of contraceptive. He is currently the Director of Surgical Research and Professor of Cell Biology at Harvard Medical School, Children's Hospital in Boston.

8.7 The angiogenic switch

The regulation of angiogenesis is dependent upon the dynamic balance of angiogenic inducers and inhibitors. Increasing the activity of the inducers or decreasing the activity of the inhibitors tips the balance of the

Anti-angiogenic factors

Proteolytic fragments
Angiostatin
Endostatin
Serpin antithrombin
Canstatin
PEX
Prolactin (16 kD)
Restin
Tumstatin
Arresten
Vasostatin
Kringle 1-5
Fibronectin fragments
Cytokines and chemokines
Interleukin -1,-4, -10, -12 and -18
Interferon -α, -β, -γ
EMAP II
gro -β
IP-10
Monokine induced by interferon -γ
Platelet factor 4
Soluble receptors
Soluble FGFR-1
Soluble VEGFR-1
Collagenase inhibitor
TIMP-1,-2,-3 and -4
Vitamins
1,25-(OH) vitamin D_3
Retinoic acid
Tumour suppresor genes
p16
p53
Other inhibitors
Angiopoietin
Angiotensin
Angiotensin-2-receptor
Caveolin
Meth-1,-2
2-Methoxy oestradiol
Osteopondin cleavage product
Pigment epithelium derived factor
Prostatic specific antigen
Protamine
Thrombospondin-1,-2
Transforming growth factor-β1
Troponin 1

Pro-angiogenic factors

Growth factors
VEGF
FGF (acidic & basic)
Hepatocyte-derived growth factor
Platelet-derived growth factor
EGF
Granulocyte colony-stimulating factor
Tumor necrosis factor α
Cytokines
Interleukin-1,-6 and -8
Enzymes
Cathepsin
Gelatinase A,B
Stromelysin
Small adhesion molecule
α_V,β_3 integrin
Metal ions
Copper
Others
Angiostatin-2
Angiopoietin-I
Angiotropin
Angiogenin
Adrenomedullin
Erythropoietin
Endothelin
Hypoxia
Midkine
Nitric oxide synthase
Prostaglandin E
Pleiotropin
Platelet activating factor
Plasminogen activator inhibitor
Thymidine phosphorylase
Thrombopoietin
Urokinase tissue plasminogen

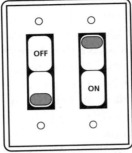

Angiogenic switch

Figure 8.5 The angiogenic switch. Adapted from Madhusudan and Harris (2002)

"angiogenic switch" to the "on" position, and vice versa (Figure 8.5, adapted from Madhusudan and Harris, 2002).

Angiogenic inducers

Growth factors, both non-specific and endothelial-specific, dominate the list of angiogenic inducers. Although the non-specific growth factors (e.g. FGF) affect many cell types, they are still important for angiogenesis. Three families of vascular-endothelium-specific growth factors and their transmembrane receptors have been identified: vascular endothelial growth factors (VEGFs) and VEGF receptors (VEGFRs), angiopoietins and Tie receptors, and ephrins and ephrin receptors. All of these receptors are tyrosine kinase receptors.

VEGF is the star player involved in the initiation of angiogenesis, while angiopoietins and ephrins are important for subsequent maturation.

PAUSE AND THINK

Could this be another target for kinase inhibitor drugs?

The VEGF family currently consists of five family members (VEGFA-E) which transmit their signal via three receptor members (VEGFR-1, VEGFR-2, and VEGFR-3). VEGFA is secreted by a range of tumor cells. The tumor microenvironment also affects the surrounding stromal cells and induction of the VEGF promoter in surrounding non-transformed cells has been demonstrated suggesting a collaboration between host and transformed cells. Also, reserves of VEGF are found in the ECM and are released by MMPs. Not only does VEGF induce endothelial cell proliferation but it also can induce permeability and leakage. This feature may be important for the initiation of angiogenesis since it has been suggested that the existing mature vessels must be destabilized before sprouting begins. VEGFR-2 mediates the endothelial effects of VEGF, while VEGFR-1 is inhibitory and VEGFR-3 is vital for lymphatic vessels. Although the details of the signal transduction pathway of VEGFRs have yet to be elucidated, it appears to be very similar to the signal transduction pathways for EGFRs (see Chapter 4): dimerization, autophosphorylation, creation of high affinity binding sites for proteins with SH2 domains, and subsequent activation of the RAS, Raf, MAP kinase cascade. A PI3 kinase dependent pathway is also implicated.

Angiogenic inhibitors

Angiogenic inhibitors normally found in the body (endogenous inhibitors) maintain the angiogenic switch in the "off" position by inhibiting endothelial cell migration and proliferation. Some angiogenic inhibitors are stored as cryptic parts within larger proteins that are not themselves inhibitors (Figure 8.6). Plasminogen can be cleaved by proteinases including several MMPs, to release the angiogenic inhibitor, angiostatin. Angiostatin binds to its endothelial cell surface receptor, annexin II to exert its inhibitory effects. Endostatin is a fragment of collagen XVIII

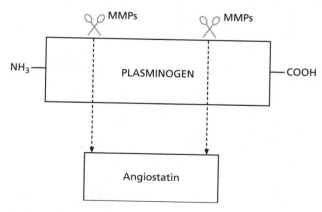

Figure 8.6 Cryptic angiogenic inhibitors

and can be proteolytically released by elastase and cathepsin. It blocks MAPK activation in endothelial cells and also MMPs.

It has been observed that sometimes, when a tumor is removed by surgery or irradiation, dormant metastases are often activated and growth and angiogenesis is initiated. This phenomenon has been termed "concomitant resistance". Evidence suggests that the production of angiogenic inhibitors, such as angiostatin and endostatin, by certain tumors prevent the growth of remote micrometastases via the blood. When the primary tumor is removed, so are these inhibitors and the angiogenic switch is activated for the micrometastases. Also, surgery is known to cause induction of angiogenic growth factors and thus may exacerbate malignant disease through this mechanism (personal communication, Ian Judson).

The angiogenic switch is regulated in two ways during tumorigenesis. First, as a tumor grows, it creates conditions of hypoxia (low oxygen concentration), and this induces angiogenesis via the hypoxia-inducible factor-1α (HIF-1α). One target of HIF-1α that is important for angiogenesis is the *VEGF* gene. HIF is actually a heterodimeric transcription factor composed of one HIF-1α and one HIF-1β subunit. The activity of HIF is regulated by oxygen concentration, not at the level of mRNA

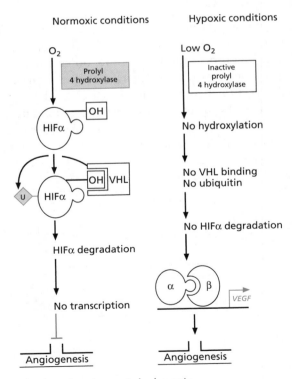

Figure 8.7 The induction of angiogenesis by hypoxia

expression as both subunit mRNAs are constitutively expressed, but rather at the protein level of HIF-1α (Figure 8.7). Under normoxic conditions (20% oxygen) HIF-1α is rapidly degraded. The von Hippel-Lindau (VHL) tumor suppressor protein is an important regulator of HIF-1α degradation (Kim and Kaelin Jr., 2003). The first step in targeting HIF-1α for degradation under normoxic conditions is modification (hydroxylation) by the enzyme prolyl 4-hydroxylase (shown in red). This enzyme directly binds and links molecular oxygen to specific proline residues on HIF-1α, and thus acts as a direct oxygen sensor in this pathway. VHL binds to hydroxylated HIF-1α and activates a complex of proteins responsible for the addition of unbiquitin (indicated by a 'U' in a red diamond, Figure 8.7) that target HIF-1α for proteosomal degradation. In the absence of HIF-1α, HIF-1α target genes cannot be transcriptional activated and angiogenesis does not occur.

Under hypoxic conditions the enzyme prolyl 4-hydroxylase is inactivated, HIF-1α is not hydroxylated, and VHL cannot bind and target HIF-1α for proteosomal degradation. HIF-1α is rapidly stabilized and transported to the nucleus. The heterodimeric HIF transcription factor can then activate its target genes. As mentioned above, the most notable target is the *VEGF* gene that contains a hypoxia response element in its promoter region.

PAUSE AND THINK

Allow me to interject at this point to share a personal story of how I have come to appreciate the importance of this pathway. I wear contact lenses and was told when I purchased them that I could sleep with them. I received a shock during the next visit with my optometrist when he told me that my eyes were not receiving enough oxygen during the night and blood vessels had begun to grow out into the eye. He explained that my eyelid alone reduces the amount of oxygen to the eye during sleep and that the addition of a contact lens created a hypoxic condition. The hypoxia was sensed and triggered angiogenesis to supply more oxygen to the eye rather then allow the tissue to become damaged. I no longer sleep with my contact lenses and have an appreciation of the regulation of prolyl 4-hydroxylase.

Secondly, oncogenic proteins and loss of tumor suppressors contribute to the modification of the angiogenic switch. In contrast to the well-known direct contribution of oncogenes and tumor suppressors to proliferation, apoptosis, and differentiation, indirect roles in angiogenesis are now recognized. Approximately 30 oncoproteins have been shown to tip the balance towards angiogenesis (Table 8.1). Aberrant growth factor production, in addition to acting in an autocrine manner to stimulate proliferation of tumor cells, can also act in a paracrine manner to stimulate the growth of endothelial cells. "Star" oncogenic proteins including receptor tyrosine kinases (e.g. EGFR), intracellular tyrosine kinases

Table 8.1 Oncogenes and altered tumor suppressor genes that are pro-angiogenic. From Kerbel and Folkman, *Nat. Rev. Cancer* 2, p. 727, Copyright (2000). Reproduced with permission

Oncogene	Mechanism of pro-angiogenic activity
Bcl-2	VEGF upregulation
EGFR	VEGF, bFGF, IL-8 upregulation
Fos	VEGF upregulation
Her2	VEGF upregulation
Jun	VEGF upregulation, thrombospondin downregulation
KRAS, HRAS	VEGF upregulation, thrombospondin downregulation
Myb	thrombospondin downregulation
Myc	angiogenic properties in epidermis
Src	VEGF upregulation, thrombospondin downregulation
Wnt	Increased VEGF
PTEN	Increased VEGF
p53	VEGF upregulation, thrombosponding downregulation
VHL	Increased VEGF
Rb	Decreased thrombospondin

(e.g. Src), intracellular transducers (e.g. Ras), and transcription factors (e.g. Fos) have been shown to upregulate the "star" angiogenic inducer, VEGF. Some of these oncogenic proteins simultaneously down-regulate angiogenic inhibitors (e.g. thrombospondin).

As one may have predicted, the multi-functional roles of the tumor suppressor p53 include the regulation of angiogenesis. As a transcription factor, p53 normally binds to and activates the promoter of the *thrombospondin-1* gene. Mutations in the *p53* gene, commonly associated with the cancer phenotype, result in a decrease of the angiogenic inhibitor so that the angiogenic switch favors angiogenesis.

8.8 Other means of tumor neovascularization

Recently, evidence suggests that. in addition to angiogenesis, vasculogenic mimicry and vasculogenesis contribute to the formation of tumor vessels (Figure 8.8a–c, respectively). Vasculogenic mimicry describes the process whereby tumor cells (e.g. melanoma cells; shown as gray circles)

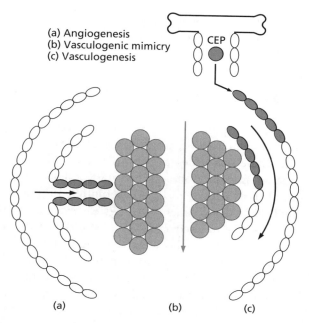

(a) Angiogenesis
(b) Vasculogenic mimicry
(c) Vasculogenesis

Figure 8.8 Tumor neovascularization

act as endothelial cells and form vascular-like structures (shown by a red arrow). Vasculogenesis involves the differentiation and proliferation of endothelial cells from endothelial progenitor cells. Studies have demonstrated up to 40% of tumor endothelial cells originated from circulating endothelial progenitor cells (CEPs; shown as a red circle) derived from the bone marrow. Angiogenic factors from the tumor, such as VEGF, are involved in the recruitment of these cells that express VEGFR-2. After reaching the tumor, CEPs differentiate and contribute to the tumor neovasculature (Figure 8.8c, red ovals). It seems likely that different cancers may differ in their requirement for CEP contributions to the new tumor vasculature. It is known that they are necessary for lymphomas and colon cancer.

◎ Therapeutic strategies

Perhaps one could envisage a therapy targeted at each of the major steps of metastasis. However, since the first and last steps are rate-limiting, these have been the most tried targets. Protease and integrin inhibitors have been obvious molecular targets to block migration. Therapies aimed at the tumor vasculature have been designed either to halt the angiogenic process (anti-angiogenic drugs) or to destroy the tumor vasculature which has been already formed (vascular targeting). Some examples of the therapies that have been developed are discussed below.

8.9 Metalloproteinase inhibitors (MPIs)

There was a rapid response by pharmaceutical companies to develop metalloproteinase inhibitors because of the evidence of the role of MMPs in metastasis. These molecules appear to function in several steps of metastasis, including migration and metastatic colonization. The initial wave of clinical trials proved to be disappointing although informative for future trials. The development of drugs exceeded the rate at which basic research was able to uncover the details of the MMP family. First, about 24 different family members have been identified. The knowledge of temporal and localization expression patterns, functional roles, and roles in different cancers of the individual family members lagged behind the development of small molecule inhibitors and natural product drugs. The trials were hindered by unexpected side effects (musculoskeletal pain) and poor design. There was difficulty in measuring efficacy and the drugs were administered only to patients with advanced disease even though preclinical evidence suggested that administration at early stages of disease was crucial. Although no MPIs have received approval as a cancer therapy thus far, modifications have been made and several MPIs are still in clinical trials. These include Marimastat; BMS-275291 (lacks named side effects); Prinomastat; Metastat; and Neovastat (isolated from shark cartilage).

8.10 Antiangiogenic therapy and vascular targeting

Anti-angiogenic therapy is designed to prevent the formation of new blood vessels. Rather than target the tumor cells directly, anti-angiogenic therapy aims at interfering with the responsiveness of endothelial cells that is essential to the tumor's survival. Drugs may be designed to prevent the cells from responding to pro-angiogenic signals or may be targeted to block the activity of the inducers (Figure 8.9). Overall, these drugs are cytostatic, rather than cytotoxic, and therefore may need long-term, continuous administration. Anti-angiogenic therapies, together with vascular targeting (discussed below), differ from the therapies discussed previously and the differences have several implications. First, since angiogenesis only occurs on occasion in the adult, drugs that inhibit it are predicted to cause minimal side effects. More importantly, the target endothelial cells recruited during angiogenesis are genetically

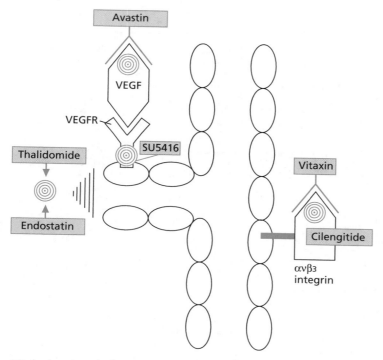

Figure 8.9 Anti-angiogenic therapies and their molecular targets

stable, unlike the tumor cells that have accumulated mutations, and therefore are less likely to develop drug resistance rapidly. Approximately 80 anti-angiogenic drugs are currently in clinical trials (some are listed in Table 8.2). Examples of several different strategies that have been employed are described below.

One strategy involves targeting angiogenic factors, such as VEGF. Avastin (Bevacizumab), a recombinant human monoclonal antibody that recognizes all VEGF isoforms, has been tested in clinical trials (for example, see Yang *et al.*, 2003) and approved for treatment of colorectal cancer. Interestingly, the drug failed to show a response (life extension) in an early breast cancer trial. Why was there a different response observed in the two different cancers? The answer lies in the angiogenic switch: colon tumors are more dependent on VEGF for the induction of angiogenesis while this is true only for early stages of breast cancer. Advanced breast cancer utilizes a broader arsenal of angiogenic inducers, and thus inhibition of just one activator does not have an effect.

Small molecule tyrosine kinase inhibitors have been used to target the VEGFR. Semaxanib (SU5416) was the first VEGFR inhibitor to enter phase III clinical trials. Its mechanism of action is that it inhibits receptor autophosphorylation. Although Semaxanib demonstrated promising results in Kaposi's sarcoma patients, significant toxicity and poor responses

Table 8.2 Angiogenesis inhibitors in clinical trials

Drug	Company	Mechanism	Phase of trial
Drugs that block activators of angiogenesis and their receptors			
SU5416	Sugen	Blocks VEGFR signaling	Withdrawn
SU6668	Sugen	Blocks VEGFR, FGFR, PDGFR	I
Avastin	Genetech	Monoclonal Ab to VEGF	Approved
IMC-1C11	Imclone System	Monoclonal Ab to VEGF	I
Angiozyme	Ribozyme Pharm	Inhibition of VEGFR synthesis	I/II
AZD2171	AstraZeneca	VEGFR 1/2 tyrosine kinase inhibitor	I
ZD6474	AstraZeneca	VEGFR 1/2 tyrosine kinase inhibitor	II
VEGF-Trap	Regeneron Pharm	Soluble decoy VEGFR	I
Drugs that inhibit endothelial-specific integrin signaling			
Vitaxin II	MedImmune	Inhibitor of $\alpha v \beta 3$ integrin	II
Cilengitide	Merck KGaA	Antagonist of integrins $\alpha v \beta 3$ and $\alpha v \beta 5$	I
Drugs that inhibit endothelial cells			
Thalidomide	Celgene	Unknown	III
Endostatin	EntreMed	Inhibition of endothelial cells	II
Angiostatin	EntreMed	Inhibition of endothelial cells	I
ABT-510	Abbott Labs	Thrombospondin-1 analog	II
Drugs that block matrix breakdown			
Marimastat	British Biotech	Inhibitor of MMPs	III
Neovastat	Aeterna	Inhibitor of MMPs	III
BMS-275291	Bristol Myers Squibb	Inhibitor of MMPs	III
Miscellaneous			
Combretastatin	Oxigene	Binds to tubulin; disrupts the cytoskeleton	I

in colorectal cancer patients led to the withdrawal of the drug. Its further development was discontinued also due to unfavorable pharmacology, namely a particularly short half-life of only several hours. As a result of the short half-life of the drug, effective doses were unable to be maintained even after bi-weekly intravenous administration. The drug, SU6668, which has a similar mode of action to Semaxanib and is orally active, is now in clinical trials. See Pause and Think, p. 175.

Administration of recombinant human endogenous inhibitors is another strategy of antiangiogenic treatment that held much promise but thus far has not delivered the expected results. Endostatin, which was only discovered in 1996, was the first to enter clinical trials. Although it was demonstrated to be non-toxic, no clinical response was observed. This negative result may, again, be due to suboptimal clinical trial design rather than inefficacy of the drug since patients with advanced solid tumors were selected despite preclinical success with early-stage cancer models. The company, EntreMed Inc. (Maryland), announced that it will halt production of endostatin due to financial difficulties.

Antagonists to integrins αvβ3 and αvβ5 would block endothelial integrin-ECM interactions and specifically induce apoptosis of angiogenic vessels with little effect on mature vessels. Two integrin inhibitors have entered clinical trials: Vitaxin is a humanized monoclonal antibody against αvβ3 and cilengitide is a synthetic cyclic peptide antagonist that mimics the Arg-Gly-Asp "ligand" sequence and inhibits integrins αvβ3, and αvβ5 (Tucker, 2002; Gutheil *et al.*, 2000).

Thalidomide, a drug cursed in the past as being teratogenic, is one of the most effective drugs for treating patients with multiple myeloma. It has been shown to inhibit angiogenesis induced by bFGF or VEGF, and reduced plasma levels of these proteins correlate with the efficacy of thalidomide treatment. However, the antiangiogenic activity of thalidomide is linked to its teratogenicity and thus patient education regarding pregnancy is crucial.

Antiangiogenic effects may be "side effects" of other cancer therapies. Cancer therapies targeted at oncogene products often affect angiogenesis. Herceptin (the antibody directed against ErbB2; see Chapter 4) has been shown to be antiangiogenic by inhibiting the production of angiogenic inducers (e.g. TGF-β, and angiopoietin-1) by tumor cells and upregulating angiogenic inhibitors (e.g. thrombospondin). Chronic frequent administration of conventional chemotherapy at doses 1/10–1/3 MTD, known as metronomic scheduling, has also resulted in antiangiogenic effects.

Vascular targeting

Vasculature targeting is a therapeutic approach designed to destroy the existing neovasculature in order to starve the tumor of oxygen and nutrients and lead to tumor regression. This approach is possible due to the identification of molecular differences between tumor and normal vasculature. Clinical trials in both the UK and USA have begun to test combretastatin compounds, first isolated from the African shrub *Combretum caffrum*, which are selectively toxic to neovasculature (Griggs *et al.*, 2001). They bind tubulin and disrupt the cytoskeleton. Its effects

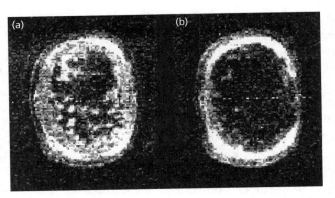

Figure 8.10 Effects of combretastatin on tumor neovasculature. Antivascular effects were analyzed by magnetic resonance imaging. Image intensity indicates tumor vasculature. A primary tumor before (a) and after (b) treatment with combretastatin A4. Strong antivascular effects are seen in the core of the tumor after treatment but a small viable rim of tumor tissue can be seen at the periphery. Reprinted from *Br. J. Cancer*, Vol. 77, Beauregard, D.A. *et al.*, "Magnetic resonance imaging and spectroscopy of combretastatin A(4) prodrug induced disruption of tumor profusion and energetic status", pp. 1761–1767, Copyright (1998), with permission by Nature Publishing Group

have been explained by the hypothesis that immature endothelium may have a more intrinsic need for a tubulin cytoskeleton to maintain their shape than stable mature vasculature which is firmly supported by a basement membrane. Loss of shape and rounding up of the endothelial cells in new blood vessels block blood flow and/or lead to vascular collapse thereby depriving the tumor of oxygen and nutrients (Figure 8.10; compare images before (a) and after (b) treatment with combretastatin A4). As a result, necrosis occurs at the core of the tumor. Unfortunately, a ring of tumor cells at the periphery remain viable (Figure 8.10b) and summon the need for combination therapy. Other mechanisms of action responsible for its apoptotic effects are also probable. Antivascular effects are seen at doses one tenth of the MTD. ZD6126 is another vascular targeting agent in clinical trials that targets the tubulin cytoskeleton. Antivascular effects have been seen without cytotoxicity (Blakey *et al.*, 2002). We must be aware for long-term administration that both drugs are selective for all neovasculature and are not only tumor specific.

Nanoparticle technology, gene therapy, and vascular targeting have come together in an exciting recent report and hints at future applications. The knowledge of several biochemical areas have been combined specifically to target an antiangiogenic gene to the neovasculature of tumors in mice (Hood *et al.*, 2002) (Figure 8.11). Lipid-based nanoparticles (shaded in red) were coated with a ligand to a neovasculature-specific receptor, integrin αvβ3. This integrin is not only specific to newly growing vessels but also internalizes viruses and small particles and thus

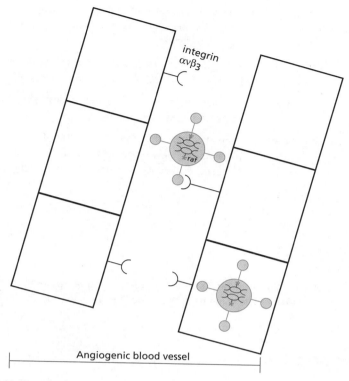

Figure 8.11 Vascular targeting by nanoparticle technology

can facilitate gene delivery. A mutant form of the *Raf* gene (marked by an asterisk) was coupled to the nanoparticles and was used to inhibit the Raf pathway critical for angiogenesis *in vivo*. Systemic delivery to mice resulted in apoptosis of tumor endothelial cells and concentric rings of apoptotic tumor cells around the targeted vessels. Regression of primary and metastatic tumors was demonstrated. Since viruses were not used for gene delivery, the disadvantages of viral delivery (e.g. risk of further carcinogenesis and an immunogenic response) were by-passed.

■ CHAPTER HIGHLIGHTS—REFRESH YOUR MEMORY

- The major steps involved in metastasis are: migration, intravasation, transport, extravasation, and metastatic colonization.

- Different cancers metastasize to specific locations due to blood flow direction and molecules which support the "seed and soil" hypothesis.

- Integrins are receptors that mediate cell–ECM interactions and, with respect to the exterior and interior of a cell, mediate bi-directional signaling.

- The steps involved in intravasation and extravasation are similar but are the reverse of each other.

- Metastatic colonization is characterized by progressive growth of a tumor at a distant site and requires angiogenesis.

- Micrometastases do not show net growth and may stay dormant for years.

- Loss of function of metastasis-suppressor genes results in an increase in metastatic capability.

- Members of the VEGF family are specific endothelial cell growth factors that are key players in angiogenesis. Their signals are mediated through transmembrane tyrosine kinase receptors.

- Hypoxia inducible factor is a heterodimeric transcription factor that targets genes important for angiogenesis, such as VEGF.

- The angiogenic switch is regulated by the dynamic balance of pro- and anti-angiogenic factors.

- Vasculogenic mimicry and vasculogenesis also contribute to neovascularization of tumors.

- Anti-angiogenic therapy is designed to *prevent the formation of* new blood vessels while vasculature targeting is designed to *destroy* the neovasculature.

■ **ACTIVITY**

Using the web sites below update Table 8.2. Have certain drugs progressed to advanced clinical trials? Have some been terminated? Have new drugs been added?

■ **FURTHER READING**

Bergers, G. and Benjamin, L.E. (2003) Tumorigenesis and the angiogenic switch. *Nature Rev. Cancer* 3:401–410.

Carmeliet, P. and Jain, R.K. (2000) Angiogenesis in cancer and other diseases. *Nature* 407:249–257.

Chang, C. and Werb, Z. (2001) The many faces of metalloproteases: cell growth, invasion, angiogenesis and metastasis. *Trends Cell Biol.* 11:S37–S43.

Coussens, L.M., Fingleton, B. and Matrisian, L.M. (2002) Matrix metalloproteinase inhibitors and cancer: trials and tribulations. *Science* 295:2387–2392.

Hood, J.D. and Cheresh, D.A. (2002) Role of integrins in cell invasion and migration. *Nature Rev. Cancer* 2:91–100.

Kerbel, R. and Folkman, J. (2002) Clinical translation of angiogenesis inhibitors. *Nature Rev. Cancer* 2:727–739.

Matter, A. (2001) Tumour angiogenesis as a therapeutic target. *DDT* 6:1005–1020.

Madhusudan, S. and Harris, A.L. (2002) Drug inhibition of angiogenesis. *Curr. Opin. Pharm.* 2:403–414.

Malik, A.K. and Gerber, H.-P. (2004) Targeting VEGF ligands and receptors in cancer. *Targets* 2:48–57.

McCarty, M.F., Liu, W., Fan, F., Parikh, A., Reimuth, N., Stoeltzing, O. and Ellis, L.M. (2003) Promises and pitfalls of anti-angiogenic therapy in clinical trials. *Trends Mol. Med.* 9:53–58.

Ruoslahti, E. (2002) Specialization of tumour vasculature. *Nature Rev. Cancer* 2:83–90.

Steeg, P.S. (2003) Metastasis suppressors alter the signal transduction of cancer cells. *Nature Rev. Cancer.* **3**:55–63.

Taraboletti, G. and Margosio, B. (2001) Antiangiogenic and antivascular therapy for cancer. *Curr. Opin. Pharm.* **1**:378–384.

▓ WEB SITES

The Angiogenesis Foundation http://www.angio.org
www.cancer.gov/clinicaltrials

▓ SELECTED SPECIAL TOPICS

Blakey, D.C., Ashton, S.E., Westwood, F.R., Walker, M. and Ryan, A.J. (2002) ZD6126: A novel small molecule vascular targeting agent. *Int. J. Rad. Oncol. Biol. Phys.* **54**:1497–1502.

Griggs, J., Metcalfe, J.C. and Hesketh, R. (2001) Targeting tumor vasculature: the development of combretastatin A. *Lancet Oncol.* **2**:82–87.

Gutheil, J.C., Campbell, T.N., Pierce, P.R., Watkins, J.D., Huse, W.D., Bodkin, D.J. and Cheresh, D.A. (2000) Targeted antiangiogenic therapy for cancer using Vitaxin: a humanized monoclonal antibody to the integrin $\alpha v \beta 3$. *Clin. Cancer. Res.* **6**:3056–3061.

Hood, J.D., Bednarski, M., Frausto, R., Guccione, S., Reisfeld, R.A., Xiang, R. and Cheresh, D.A. (2002) Tumor regression by targeted gene delivery to the neovasculature. *Science* **296**:2404–2407.

Kim, W. and Kaelin Jr., W.G. (2003) The von-Hippel-Lindau tumor repressor protein: new insights into oxygen sensing and cancer. *Curr. Opin. Gen. Dev.* **13**:55–60.

Laferriere, J. Houle, F. and Huot, J. (2002) Regulation of the metastatic process by E-selectin and stress-activated protein kinase-2/p38. *Ann. NY Acad. Sci.* **973**:562–572.

Tucker, G.C. (2002) Inhibitors of integrins. *Curr. Opin. Pharm.* **2**:394–402.

Yang, J.C., Haworth, L., Sherry, R.M., Hwu, P., Schwartzentruber, D.J., Topalian, S.L., Steinberg, S.M., Chen, H.X. and Rosenberg, S.A. (2003) A randomized trial of bevacizumab, an anti-vascular endothelial growth factor antibody for metastatic renal cancer. *New Engl. J. Med.* **349**:427–434.

9

Nutrients, hormones, and gene interactions

Introduction

Does our diet influence whether we are the one out of three people who get cancer? Diet plays a significant role in cancer incidence. Many epidemiological studies provide evidence to support the role of diet in both causation and prevention of cancer. Approximately one third of the variations in cancer incidence between different populations is due to differences in diet. For example, the Japanese diet has changed radically between the 1950s and 1990s including a seven-fold increase in meat consumption (Key *et al.*, 2002). This coincides with a five-fold increase in colorectal cancer over the same period. In this chapter we will see how nutrients act as causative and preventative factors. The knowledge gained from investigations into the role of diet in cancer and cancer prevention should be integrated into lifestyle modifications in order to reduce the occurrence of the disease. The mechanisms of action of nutrients will also be examined and we will see that some parallel the mechanism of action of hormones. The chapter will conclude with a discussion of the role of hormones in carcinogenesis.

Let us examine why we eat food (Figure 9.1). The basic food groups of carbohydrates, fats, and proteins provide us with glucose, fatty acids, and amino acids that can be metabolized to produce energy. Food also provides precursors for biosynthetic reactions. Proteins provide a source of nitrogen needed for the synthesis of the nitrogenous bases of DNA. Cofactors provided by vitamins and minerals in our diet are essential for the function of many enzymes. Additional biologically active micro-constituents have been identified in the foods we eat (Table 9.1). Very recently, some nutrients and micro-constituents have been shown to affect gene expression. Many biologically active micro-constituents act as antioxidants, compounds that significantly inhibit or delay the damaging action of reactive oxygen species (see Chapter 2), often by being oxidized themselves. Carotenoids and polyphenols, groups of plant metabolites that are found in many fruits, vegetables, and teas, are examples. Plants require many phytochemicals as a defense against excess energy and oxidative damage as they absorb solar energy for photosynthesis. Many of these phytochemicals also protect human cells. Some antioxidants required by humans can be synthesized in the body but others must be obtained by eating fruit and vegetables. The four major groups of

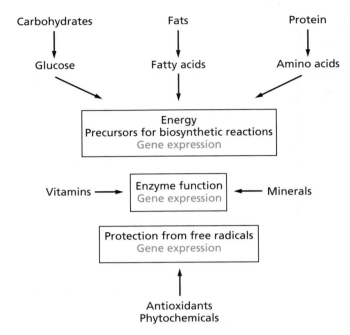

Figure 9.1 Provisions of food

dietary antioxidants-phytochemicals are vitamin C, isoprenoids (e.g. vitamin E), phenolic compounds (flavonoids), and organosulfur compounds. All four will be discussed later in the chapter.

Information gained about the role of micro-constituents could be applied to the development of chemopreventative supplements, extra sources of dietary components taken in addition to food. However, unraveling the individual contributions of micro-constituents as preventative agents against cancer is a challenge for the future. Epidemiological studies strongly suggested diets rich in β-carotene-containing fruits and vegetables reduced lung cancer risk. Animal studies also generated supportive evidence. This led to the The β-Carotene and Retinol Efficacy Trial (CARET) and the Alpha-Tocopherol Beta-Carotene Cancer Prevention Study (ATBC) which tested the effect of β-carotene supplements on smokers and those exposed to asbestos. Surprisingly, β-carotene supplementation increased lung cancer in these high-risk individuals and had no effect on healthy individuals. These trials are examples of studies formulated on preclinical findings that generate data that do not support an initial hypothesis. See Pause and Think.

Only recently have molecular approaches been used to investigate the molecular mechanisms of dietary constituents involved in the causation or prevention of cancer. One of the most significant insights gained is that **nutrients regulate gene expression**. The power of food has begun to be revealed. This chapter will include a sample of these findings.

PAUSE AND THINK

Propose a hypothesis to explain the seemingly conflicting results. The most likely explanation is that alternative micro-constituents in β-carotene rich vegetables and fruits may be the active ingredient in reducing lung cancer risk, or perhaps β-carotene works in a synergistic manner with other micro-constituents not present in the supplements. Interactions between different dietary constituents must be considered for a complete picture.

Table 9.1 Microconstituents. Reprinted from *Trends Mol. Med.*, Manson, M.M., "Cancer prevention—the potential for diet to modulate molecular signaling", Vol. 9 pp. 11–18, Copyright (2003). Reprinted with permission from Elsevier

Food source	Class of compound	Chemical
Cruciferous vegetables	Isothiocyanate	Benzyl isothiocyanate, phenethyl isothiocyanate, sulforaphane
Cruciferous vegetables	Dithiolthione	Ohipraz[2]
Cruciferous vegetables	Glycosinolate	Indole-3-carbinol, 3,3'-diindoylmethane, indole-3-acetonitrile
Onions, garlic, scallions, chives	Allium compound	Diallyl sulphide
		Allylmethyl trisulphide
Citrus fruit (peel)	Terpenoid	D-Limonene, penllyl alcohol, geraniol, menthol, carvone
Citrus fruit	Flavonoid	Tangeretin, nobiletin, ratin
Berries, tomatoes, potatoes, broad beans, broccoli, squash, onions	Flavonoid	Quercetin
Radish, horse radish, kale, endive	Flavonoid	Kaempferol
Tea, chocolate	Polyphenol	Epigallocatechin gallate, epigallocatechin, epicatechin, catechin
Grapes	Polyphenol	Resveratrol
Turmeric	Polyphenol	Curcumin
Strawberries, raspberries, blackberries, walnuts, pecans	Polyphenol	Caffeic acid, ferulic acid, ellagic acid
Cereals, pulses (millet, sorghum, soya beans)	Isoflavone	Genistein
Orange vegetables and fruit	Carotenoid	α- and β-carotene
Tomatoes	Carotenoid	Lycopene
Tea, coffee, cola, cacao (cocoa and chocolate)	Methylxanthines	Caffeine, theophylline, theobromine

9.1 Causative factors

Three main aspects of our diet can be considered as causative factors of cancer. First, any given food is a very complex substance that can carry harmful factors, in addition to nutritional value. The consumption of food provides a route for chemical carcinogens to be delivered to the body. Genotoxic agents present as micro-constituents in food act as dietary carcinogens. Secondly, lack of a particular essential nutrient may enhance cancer risk. In addition, conditions such as obesity, aid in tumor promotion. In this section we will examine carcinogenic contaminants, nutritional deficiencies, and obesity, as dietary cancer-causative factors.

Carcinogenic contaminants

The carcinogenic effect of one particular food can be variable. Salmon, rich in omega-3 polyunsaturated fatty acids and known to be an important component of a healthy diet, is one example. Salmon, being fatty carnivorous fish, accumulate pollutants and can pass genotoxic contaminants through the food chain to humans. A study of farmed and wild salmon from around the world found that in some geographical regions (e.g. Scotland), polychlorinated biphenols (PCBs) and other pesticides are present in quantities that suggested that eating farmed salmon more than once a month could increase cancer risk (Hites *et al.*, 2004). Risk was calculated based on the assumption that the risks of individual carcinogens are additive. This study raises many issues. First, it underscores that differences in the source of food can have varying consequences; farmed salmon has more contaminants than wild salmon and farmed salmon from Scotland contains significantly more contaminants than farmed salmon available in North American cities. Perhaps the results have a broader implication and point to the suggestion that the source of all food should be properly labeled to allow for consumer choice and to create competition for the production of good products. Data from this study suggested that fish feed (fish meal and fish oils) may be a distinguishing factor for the carcinogenicity of salmon and suggests that improvements in feed composition are needed. Evaluating cancer risk associated with more than one contaminant at a time, in addition to the benefits of other micro-constituents of a particular food, is an area that requires further study. Overall, the study described above underlines the complications that occur when analyzing the relationship between diet and cancer.

Food preparation can contribute to the cancer-causing properties of our diet. Heterocyclic amines produced by cooking meat at high temperatures have been discussed as carcinogens in Chapter 2. After metabolic activation, their mechanism of action involves the formation of DNA-adducts, resulting in base substitutions and thus mutations. Similarly, toxins produced by molds that contaminate food form DNA-adducts and thus are genotoxic. Aflatoxin B, a fungal product of *Aspergillus flavus*, is a well-known contaminant found on peanuts, and Fumonisin

PAUSE AND THINK

Suggest an experiment that examines whether the nutritional benefit of a particular food outweighs its risk as a carcinogen.

B is found on corn. Aflatoxin induces GC to TA transversions and is thought to be involved in hepatocellular carcinoma. Food preservatives, such as sodium nitrite, are regulated by government agencies because they, too, are a risk factor producing carcinogenic N-nitroso compounds.

Dietary deficiencies

Evidence is accumulating that supports the concept that micronutrient deficiencies also contribute to cancer risk. The most compelling findings suggest that the body's folate level affects colorectal cancer. Folate, one of the B vitamins, can accept or donate one-carbon units in metabolic reactions. Folate is a critical coenzyme for nucleotide synthesis and DNA methylation and these processes can affect carcinogenesis. The enzyme, methylenetetrahydrofolate reductase (MTHFR) regulates the balance between nucleotide synthesis and DNA methylation by affecting the relative quantities of 5,10-methylenetetrahydrofolate (5,10-methylene THF) and methyl-tetrahydrofolate (5-methyl THF), the precursors of these distinct processes, respectively (Figure 9.2a). MTHFR irreversibly converts 5,10-methylene THF to 5-methyl THF. 5,10-methylene THF and deoxyuridylate(dUMP) are reactants for the enzyme thymidylate

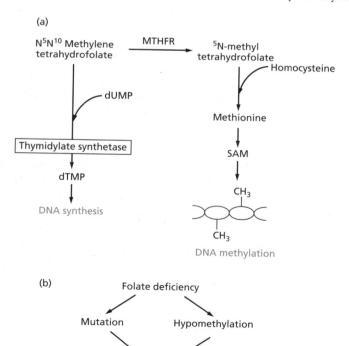

Figure 9.2 (a) Role of folate derivatives in DNA synthesis and DNA methylation; (b) Role of folate deficiency in cancer

synthase used for the production of deoxythymidylate(dTMP). 5-methyl THF and homocysteine are reactants used to produce methionine, which regenerates S-adenosylmethionine (SAM), the methyl donor for DNA methylation.

The depletion of folate may contribute to tumor development by interfering with both nucleotide synthesis and DNA methylation. A disruption in DNA synthesis leads to DNA instability and fuels mutation while disruption in DNA methylation also fuels carcinogenesis (Figure 9.2b). Deoxythymidylate synthesis is inhibited in conditions of low folate and the imbalance of the nucleotides results in the incorporation of uracil into DNA. DNA strand breaks occur as a result of attempts to repair this DNA and these breaks increase cancer risk. Both uracil misincorporation and DNA strand breaks are observed in folate-deficient humans and both defects are reversed by folate administration. Remember that genomic hypomethylation and specific tumor suppressor gene promoter hypermethylation is characteristic of the epigenetic changes observed in cancer cells (Chapter 3). Since the methyl groups used for DNA methylation are supplied by folate, a lack of folate causes a decrease in the synthesis of methionine, and subsequently genomic hypomethylation of DNA. Genomic hypomethylation is observed in folate-deficient humans and is reversed upon folate repletion. Hypermethylation at specific $5'$-gene loci have also been observed during studies of folate depletion.

Obesity

Obesity, classified as a risk factor for several cancers, is the excessive accumulation of fat that leads to a body weight that is beyond the limitation of skeletal and physical requirements. Those with a body mass index (weight/height squared) greater than 30 kg/m^2 are considered obese. It has become a significant problem in the USA, affecting 25% of the population. Several mechanisms of action of obesity as a cancer risk factor have been suggested. First, obesity is associated with acid reflux, which damages the esophageal epithelium and leads to adenocarcinoma of the esophagus. Secondly, obesity results in high fat deposits in adipose cells. The deposits may be used for the synthesis of estrogen from androgen by aromatase and may contribute to breast cancer risk (see below). Thirdly, food metabolism is linked with oxidation and an increase in ROS production can cause an increase in mutations.

9.2 Preventative factors

Fruits and vegetables

The intake of fruits and vegetables as a means of reducing cancer risk is strongly supported by epidemiological studies. Ongoing studies, such as the European Prospective Investigation into Cancer and Nutrition (EPIC), include the collection of blood samples for analysis and this will provide a valuable source of chemical and molecular data for future studies. The ability to block DNA damage caused by reactive oxygen species and/or carcinogens is the most direct strategy for preventing the initiation of cancer and for slowing down disease progression. It is here that nutrients play an important role. This is accomplished either directly by free radical scavengers (below), or indirectly by regulating the activity of Phase I and Phase II metabolizing enzymes in the body. The modulation of these metabolizing enzymes is a major defense mechanism against xenobiotics (foreign substances). The cytochrome P450 family of Phase I drug metabolizing enzymes catalyzes the hydroxylation of many drugs which makes them more water soluble for excretion but, at the same time, often has a harmful effect by converting procarcinogenic molecules into ultimate carcinogens. Phase II enzymes, such as UDP-glucuronosyltransferases or glutathione S-transferases, catalyze conjugation reactions, thus modifying xenobiotic compounds and aiding their removal from the cell.

Figure 9.3 illustrates the modifications of the carcinogen benzo[a]pyrene made by the Phase I and II metabolizing enzymes. First, it undergoes hydroxylation by cytochrome P450 and then 3-hydroxybenzo[a]pyrene is conjugated to UDP-glucuronate to produce the water-soluble and easily excreted molecule, hydroxybenzo[a]pyrene glucuroniside.

Free radical scavenging

Several micro-constituents in fruits and vegetables act as antioxidants that scavenge ROS. Water-soluble vitamin C (Figure 9.4a) can donate an electron to a free radical directly, thus inhibiting its reactivity and blocking free radical chain reactions. Oxidized vitamin C forms an ascorbyl radical that is fairly stable and unreactive due to electron delocalization or resonance. An enzyme called vitamin C reductase can regenerate vitamin C from the ascorbyl radical for reuse, or the ascorbyl radical may lose another electron and become degraded. Consequently, vitamin C reserves need to replenished daily. Lipid-soluble vitamin E (Figure 9.4b) acts as a free radical scavenger in a similar manner. A resonance-stabilized structure called the α-tocopheryl radical is produced after vitamin E donates an electron to a free radical (e.g. singlet oxygen) and helps to terminate free radical chain reactions in membranes.

Nutrient–gene interactions

"You are what you eat" is a common expression. It has been given greater significance recently by its translation into molecular terms: some

PAUSE AND THINK

In general, how would you modify Phase I and II enzyme activity to reduce tumor formation? Inhibit Phase I and induce Phase II.

PAUSE AND THINK

Is the hydroxyl radical described in Chapter 2 likely to be scavenged by these micro-constituents? No, in fact the menacing reactive hydroxyl radical is unlikely to be scavenged by these micro-constituents because of its extremely rapid reaction time. It would take exceptionally high concentrations to prevent such interactions with molecules immediately surrounding it.

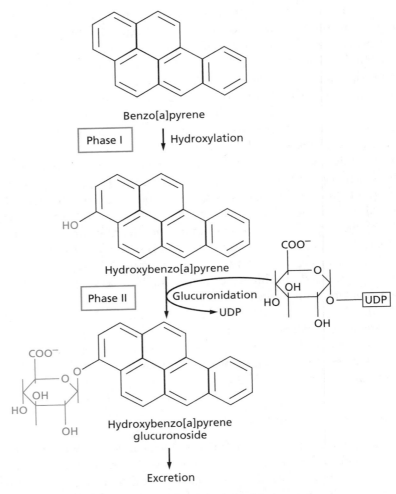

Benzo[a]pyrene

Phase I   ↓ Hydroxylation

HO   Hydroxybenzo[a]pyrene

Phase II   Glucuronidation

→ UDP

Hydroxybenzo[a]pyrene
glucuronoside

↓

Excretion

Figure 9.3 Effects of Phase I and II enzymes on benzo[a]pyrene

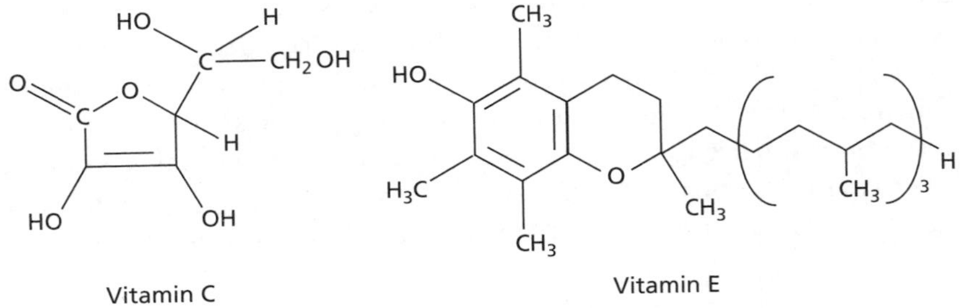

Vitamin C   Vitamin E

Figure 9.4 The structure of the antioxidant vitamins, vitamin C and vitamin E

dietary constituents can affect the expression of our genes. The molecular mechanisms employed are common to those discussed in previous chapters. For examples, dietary constituents such as isothiocyanates and polyphenols found in vegetables and fruits (e.g. broccoli and grapes), activate MAPK signal transduction pathways (Figure 9.5). As with the induction of this pathway by growth factors (Chapter 4), kinases modulate the activity of transcription factors by phosphorylation and the end result is the regulation of gene expression. The crucial link between nutrients and their role in preventing DNA damage was made by the identification of an antioxidant response element (ARE) (5′ A/G TGA C/T NNNGC A/G-3′) in the promoter region of the genes for several drug metabolizing and antioxidant enzymes (e.g. glutathione-S-transferase). The ARE conferred antioxidant-dependent regulation of target genes. That is, genes containing an ARE in their promoter regions are transcriptionally activated in response to antioxidants. The transcription factors Nrf2 and Maf, members of the basic leucine zipper family, bind the ARE and are candidates for mediating the effects of the MAPK signal transduction pathway. Therefore, antioxidants can exert some effects by regulating the expression of genes that code for detoxifying enzymes.

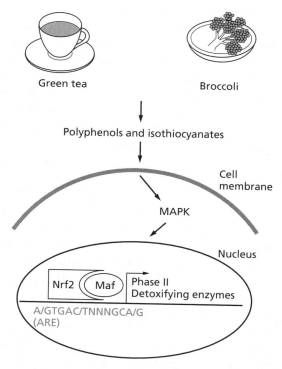

Figure 9.5 Dietary constituents regulate gene expression via MAP kinase

LIFESTYLE TIP

We should use the knowledge we have gained about the preventative role of particular foods and beverages to make better choices about what we ingest. Green tea is a better choice in comparison to soda.

One recent study examined the molecular effects of fruit polyphenols in humans by analyzing DNA damage (Bub *et al.*, 2003). The effects of fruit juice consumption (330 ml/day; over two, 2-week periods) in healthy men were studied. Two juices were tested; both contained apple, mango, and orange juice but one was enhanced with berries rich in anthocyanin and the other with green tea, apricot, and lime rich in flavanols. Using a specific endonuclease to detect oxidized pyrimidine bases, the data showed significantly lower levels of DNA base oxidation for both juices. The effect was observed after the second 2-week period of consumption, but not the first. It was not permanent since levels returned to baseline when tested 11 weeks after the experiment was terminated. The time delay indicated by this experiment suggests that ROS scavenging is not the prime mechanism and that protective detoxifying enzymes (described above) are induced.

Additional mechanisms of dietary micro-constituents

Current evidence suggests several mechanisms for the cancer-preventative role of fruits and vegetables. As we have seen above, one mechanism is the ability to decrease oxidative DNA damage by free radical scavenging or inducing protective enzymes. Two other mechanisms for the role of particular vegetables in cancer prevention are modulation of apoptosis and/or cell proliferation. The micro-constituents of garlic utilize all three mechanisms. The antioxidant properties of organosulfur compounds in garlic include the induction of Phase II enzymes and scavenging. Ajoene, a major compound in garlic, has been shown to induce apoptosis of leukemic cells in patients with leukemia. Particular caspases (3 and 8) and transcription factors (κB) are activated and peroxide is produced. Allicin, another major compound in garlic, has been shown to inhibit the proliferation of human mammary, endometrial, and colon cancer cells.

Green tea is the second most popular beverage in the world, after water. High intake of green tea is associated with low incidence of several cancers (e.g. gastric and colorectal cancer). Upon ingestion, the major polyphenol of green tea, epigallocatechin gallate (EGCG), undergoes rapid degradation and by an unknown mechanism blocks telomerase activity (Naasani *et al.*, 2003). The inhibition of telomerase limits the replicative capacity of cells (see Chapter 3) and in this study correlates with a decrease in tumor size in mouse models.

Although fiber is usually included in discussions of preventative agents of cancer, I have chosen to omit this topic from this discussion due to the inconsistencies of recent large studies (see references within Key *et al.*, 2002). On the other hand, the first key publication of EPIC shows a strong protective effect of fiber in food against colorectal cancer (Bingham *et al.*, 2003). The data suggests that preventative effects were not seen in some previous studies because the range of fiber intake was much lower than those in the EPIC study.

In conclusion, a brief examination of several different foods demonstrates that the molecular mechanisms by which nutrients affect carcinogenesis are beginning to be revealed.

9.3 Genetic polymorphisms and diet

It seems that some people can do all the "wrong things" such as drink excessive amounts of alcohol and smoke heavily and still live long healthy lives. We often hear that this is due to an individual's metabolism. Cancer risk associated with diet is influenced by an individual's metabolism. Metabolic reactions are catalyzed by enzymes. Enzyme activities may vary among individuals due to small variations, often single nucleotide changes, in the genes that code for them. These genetic polymorphisms may alter the response to a particular dietary constituent. Here are two examples. A polymorphism (C to T transition at nucleotide 677) in the MTHFR gene (see Section 9.1) reduces its enzyme activity and homozygotes for this polymorphism have a 50% decreased risk in colorectal cancer compared with those with wild-type alleles (Ueland *et al.*, 2001). These individuals have an increased availability of 5,10-methylene THF and a lower chance of disrupting nucleotide and subsequently DNA synthesis; these conditions deter mutation and carcinogenesis. However, the polymorphism increases the risk of cancer if these individuals become deficient in folate. Under these conditions, methyl-THF becomes depleted and DNA methylation is altered in a manner that is characteristic of carcinogenesis.

Polymorphisms in the gene that codes for N-acetyltransferase modify the risk of specific cancers in response to the consumption of red meat. This enzyme is involved in the metabolic activation of carcinogenic heterocyclic amines produced by cooking meat at high temperatures. Individuals with the "rapid variant" of the enzyme (a fast acetylator) who consume large amounts of red meat have an increased risk of colon cancer compared with those who have this variant and do not consume much red meat, or those who possess the "slow variant" polymorphism that do. Therefore, the response to red meat intake with respect to an increased risk of cancer depends on a person's genotype in combination with exposure to carcinogens that result from cooking.

Inherited metabolic diseases can illustrate a more obvious role of metabolism in carcinogenesis. Here are two examples resulting from blocks in tyrosine metabolism pathways. Albinos have an inherited deficiency of the enzyme tyrosinase and are unable to produce melanin causing the characteristic lack of pigment in their skin. The lack of pigment causes albinos to be more sensitive to the sun and results in an increased risk of skin carcinoma. Tyrosinemia type I, another disorder of tyrosine metabolism, results from a deficiency of fumarylacetoacetate hydrolase. As a result of this metabolic block, the metabolites fumarylacetoacetate and maleylacetate accumulate. Both are alkylating agents and cause DNA mutations and tumorigenesis. In brief, tyrosinemia type I is characterized by the synthesis and accumulation of carcinogens.

9.4 Vitamin D: a link between nutrients and hormone action

The link between nutrients and molecular signaling became apparent upon the discovery that the receptors for vitamin A and D are members of the steroid hormone receptor superfamily.

Let us examine the molecular mechanisms of vitamin D action. Vitamin D can be obtained through the diet (dairy products and seafood) or produced in the skin from 7-dehydrocholesterol upon exposure to sunlight. Its active form, $1\alpha,25$-hydroxyvitamin D_3, acts as a ligand for the cytoplasmic vitamin D receptor, a member of the steroid hormone receptor superfamily. This receptor recognizes the vitamin D response element in gene promoter regions and regulates transcription of its target genes. Current data suggest that vitamin D is a chemopreventive agent that inhibits growth and induces differentiation and apoptosis through several molecular targets. Here are a few examples. Vitamin D can act as a dominant negative ligand for EGFR (epidermal growth factor receptor, Chapter 4). That is, vitamin D can bind to the ligand binding domain of EGFR instead of EGF and prevent the binding of EGF to EGFR. As a result, vitamin D can inhibit growth. Secondly, upon binding to the vitamin D receptor, the active form of vitamin D can directly activate specific tumor suppressor genes such as the BRCA-1 gene and the p21 gene through a vitamin D response element in their promoter regions. You may recall that the p21 protein is an inhibitor of cyclin dependent kinase and that it can induce cell cycle arrest. Vitamin D promotes apoptosis through mitochondrial signaling independent of caspase activation. It induces the redistribution of two pro-apoptotic proteins BAK and BAX from the cytosol to the mitochondria. These two proteins form channels in the mitochondrial membrane and facilitate cytochrome c release and apoptosome assembly (see Chapter 6). Simultaneously, Bcl-2 and IAPs, inhibitors of apoptosis, are down-regulated.

PAUSE AND THINK

How do steroid hormone receptors function? Remember from Chapter 3 that they are ligand dependent transcription factors.

9.5 Hormones and cancer

There is group of cancers linked by a common mechanism of carcinogenesis that involves endogenous hormones as initiators rather than chemicals, viruses, or radiation. The hormone-related cancers include breast, endometrium, ovary, prostate, testis, and thyroid cancer. We will examine breast cancer as a paradigm for hormonal carcinogenesis. (Note: links between hormones and some other hormone-related cancers are not as straightforward.)

Breast cancer is the most common type of cancer among women. Estrogens (estradiol and estrone) appear to play a central role in the initiation and progression of breast cancer. The principal site of estrogen synthesis in the body changes with increased age: the ovaries are the main source in premenopausal women and adipose tissue (fat) is the main source in post-menopausal women. Life events, such as pregnancy, affect the exposure time to estrogens and those events that prolong exposure are considered risk factors for breast cancer (Figure 9.6). Early menarche (the start of the menstrual cycles) and late menopause indicate an extended length of time when the ovaries are producing estrogen. Obesity in post-menopausal women is a risk factor because adipose cells are the primary source of estrogen at this stage in life. Obese women have an increased number of fat cells and therefore produce increased amounts of estrogen. Adipose cells use the enzyme aromatase to produce estrogen from androgens. Thus, obesity increases breast cancer risk through increased estrogen production. It is thought that alcohol consumption increases breast cancer risk through a similar mechanism of increased estrogen production. Some data suggest that one alcoholic drink per day causes a 7% increase in breast cancer risk. The use of exogenous hormones in oral contraceptives and in hormone replacement therapy have also been linked to increased risk of breast cancer. Conversely, factors that interrupt the menstrual cycle such as pregnancy, lactation, and physical activity are considered protective factors. Although a rare event, breast cancer can occur in men. Men also produce some amounts of estrogen and those with high levels appear to be more at risk.

Two predominant models for the mechanisms by which estrogens exert their effects have been proposed (Figure 9.7). It is likely that both mechanisms contribute to breast cancer.

One model is that estrogens promote cell proliferation of the breast and the high division rate allows less time for DNA repair thus creating an opportunity for errors to occur during DNA replication. The increase in error rate translates into an increase in somatic mutations that lead to carcinogenesis. This model contrasts chemical and radiation-induced carcinogenesis, since no specific initiator other than errors in replication is required. Estrogen does indeed act as a mitogen for cells in the

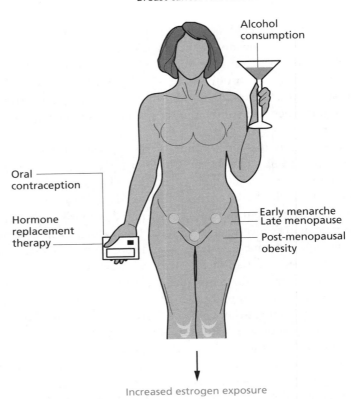

Figure 9.6 Breast cancer risk factors

breast that contain estrogen receptors. During pregnancy, estrogen levels increase and cause the mammary ducts to grow and the breasts to nearly double in size. Effects of estrogen are mediated through estrogen receptors (ERs), estrogen receptor-α and estrogen receptor-β, members of the steroid hormone receptor superfamily (see Chapter 3). The expression of the estrogen receptor isoforms changes during carcinogenesis of the breast. Estrogen receptor-α is significantly up-regulated and estrogen receptor-β is down-regulated in the majority of breast cancers. The reasons for these alterations are presently unknown. However, as we shall see below, blocking ER function has proved to be a successful strategy for the treatment of breast cancer.

The involvement of estrogen signaling in increasing breast cancer susceptibility is supported by studies of the Breast Cancer Susceptibility Genes (BRCAs). About 5–10% of all cases of breast cancer are due to an inherited predisposition, 85% of which result from germline mutations in either the *BRCA1* or *BRCA2* gene. The *BRCA* gene products are nuclear tumor suppressor proteins that play a role in transcriptional

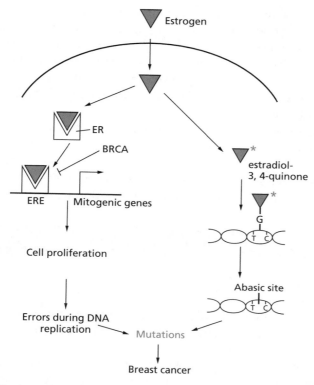

Figure 9.7 Carcinogenic mechanisms of estrogens

regulation, DNA repair, and regulation of the cell cycle. BRCA1 inhibits the transcriptional activation activity of estrogen receptors and this suggests that it suppresses the proliferating effects of estrogen signaling. It is thought that patients with an inherited mutation of *BRCA1* have a predisposition to breast cancer because they have lost modulation of estrogen signaling.

Another model suggests that estrogen and its metabolites are genotoxic. This model is consistent with the mechanism by which chemicals, viruses, and radiation initiate carcinogenesis. Estradiol is metabolized to form estradiol-3,4-quinone in cells. This metabolite covalently binds to adenine or guanine bases. The resulting adducts destabilize the bonds linking the base to the DNA backbone and result in an abasic site and, ultimately, mutations. [Note: since adenine and guanine are purines, abasic sites that involve the loss of adenine or guanine are referred to as apurinic sites.] Estradiol quinones are present in human breast tissue. Estrogen receptor knockout mice were used as an animal model to test the effect of estrogen in the absence of estrogen receptors. Results demonstrated that genotoxic effects occurred in the absence of estrogen receptors (Yue *et al.*, 2003). Data obtained from human breast epithelial cells *in vitro* indicated that metabolites of estrogen induce transformation *in vitro* and also induce loss of heterozygosity at chromosomal regions that have been reported affected in primary breast tumors (Russo *et al.*, 2003). These genotypic changes were not blocked by inhibitors of the estrogen receptor and showed that this effect was not receptor-mediated. Together, these results support the concept that estrogen metabolites contribute to breast carcinogenesis.

For this second model, predispositions to breast cancer may involve germline mutations in genes involved in estrogen biosynthesis and metabolism. Loss of the BRCA tumor suppressor proteins may leave breast cells more susceptible to the genotoxic effects of estrogen metabolites due to the normal role of BRCA tumor suppressor proteins in DNA repair.

◎ Therapeutic strategies

9.6 'Enhanced' foods

Foods are not yet considered therapeutic and most are not considered as preventative agents against cancer. However, as we learn more about the important role of nutrients in cancer and as our skills for manipulating food composition increases, this concept is set to change. The development of enhanced food products (food that have altered levels of particular micro-constituents), derived in some cases from genetically modified crops, will begin to flood the market. Foods have already been produced to have increased levels of antioxidants and may be used in future chemopreventative diets. Tomatoes have been classically bred to be bright red for consumer appeal and as a result contain more lycopene. Tomatoes have also been engineered to contain increased levels of zeaxanthin by overexpressing enzymes utilized in its synthesis. Recently, broccoli, containing high levels of glucosinolates (which are hydrolyzed to isothiocyanates; discussed in Section 9.2), has been developed by a traditional plant-breeding program and licensed to Seminis Inc., the world's largest developer and grower of vegetable and fruit seeds. A balance of potential advantage from enhanced foods must be tempered by the likelihood that we may have adverse reactions to food constituents at abnormally high concentrations.

9.7 Drugs that target estrogen

There are two strategies for the design of drugs that target estrogen action (Figure 9.8). The first is to design drugs that antagonize the actions of estrogens by interacting with estrogen receptors in order to

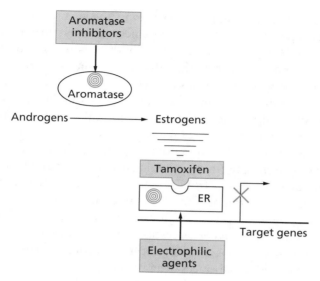

Figure 9.8 Drugs that target estrogen action

block growth in 'estrogen-positive' tumors. These compounds are called antiestrogens. Tamoxifen, used in the clinic for over 30 years, is the most widely used antiestrogen to treat estrogen receptor positive, post-menopausal breast cancer. Tamoxifen is a competitive inhibitor that alters the folding of the ligand binding domain of the estrogen receptor and blocks its ability to transactivate and initiate transcription of its target genes. Recently, the principle of a potentially new strategy for interfering with estrogen receptor function has been tested. Several electrophilic agents have been shown to target the estrogen receptor zinc finger domain and selectively block the receptor's DNA binding activity. Furthermore, breast carcinoma progression was inhibited in mice models by these drugs (Wang *et al.*, 2004). These drugs may act as lead compounds for the development of new breast cancer therapies.

The second strategy for blocking the function of estrogen in breast cancer is to design drugs that interfere with estrogen synthesis. These drugs target aromatase, the enzyme that converts androgens into estrogens. Targeting this enzyme does not interfere with the synthesis of other steroids but is rate-limiting for estrogen production. Aromatase is the main source of estrogen production in post-menopausal women because the ovaries no longer provide this function. Aromatase inhibitors can be prescribed to reduce the level of estrogen in the body of post-menopausal breast cancer patients, making it appropriate for over 50% (estrogen positive and post-menopausal) of all breast cancers. These drugs are not an option for pre-menopausal patients since aromatase is not the major estrogen producer. See Pause and Think.

PAUSE AND THINK

How would you design an aromatase inhibitor? One way is to model your compound after androstenedione, the endogenous substrate for the enzyme. These drugs should interact and inactivate the steroid-binding domain of aromatase.

Three aromatase inhibitors have gone through clinical trials and have been approved in the USA: exemestane, anastrozole, and letrozole (Brodie, 2002). The aromatase inhibitor anastrozole (Arimidex) has been used as both a first-line therapy and an adjuvant (treatment given after surgery to prevent recurrence). Results from a large-scale trial called the ATAC (Arimidex, Tamoxifen Alone or in Combination) adjuvant breast cancer trial has yielded encouraging results for the aromatase inhibitor, showing a 50% greater reduction of contralateral breast cancer at three years compared with tamoxifen. Further observation and evaluation for long-term side effects including osteoporosis and bone fractures will be needed. The result from the ATAC trial is to be expected when considering the two proposed mechanisms of action of estrogen. Tamoxifen blocks estrogen receptor mediated effects only. Aromatase inhibitors reduce total estrogen concentrations and therefore block both estrogen receptor mechanisms and non-receptor mediated genotoxic events. Molecular knowledge advances drug design.

■ **CHAPTER HIGHLIGHTS—REFRESH YOUR MEMORY**

- Diet plays a role in both the causation and prevention of cancer.

- In contrast to the association of a β-carotene rich diet with reduced lung cancer incidence, β-carotene supplementation increases lung cancer in smokers.

- Diet contributes to carcinogenesis by the delivery of carcinogenic contaminants.

- Obesity and lack of specific nutrients are also causative factors in cancer.

- Folate deficiency affects nucleotide synthesis and DNA methylation.

- Dietary constituents can regulate gene expression.

- The antioxidant response element (ARE) is found in the gene promoters of detoxifying and antioxidant enzymes.

- Nutrients may work as chemopreventative agents by blocking DNA damage via scavenging or the induction of metabolizing enzymes; inducing apoptosis; or inhibiting cell proliferation.

- The major polyphenol of green tea, epigallocatechin gallate, inhibits telomerase.

- Genetic polymorphisms can interact with nutrient status and affect cancer risk.

- Vitamin D is a nutrient that acts through a member of the steroid hormone receptor family.

- Breast cancer is a paradigm for hormonal carcinogenesis.

- Estrogen acts as a mitogen for cells in the breast.

- Estrogen and its metabolites may damage DNA directly to initiate carcinogenesis.

- Dietary factors such as obesity and alcohol increase breast cancer risk by increasing estrogen production.

- Germline mutations in the BRCA1/2 genes predispose patients to breast cancer.

- Tamoxifen is a breast cancer drug that acts as an "anti-estrogen" and blocks estrogen binding to its receptor.

- Aromatase inhibitors, such as anastrozole, act by inhibiting the enzyme aromatase that converts androgens to estrogen.

■ ACTIVITY

(i)   Record what you ate and drank for dinner last night. Give a critical account of how the meal contributed either to reducing or enhancing your cancer risk. Bon appetit!

(ii) Review the literature and using experimental evidence critically discuss the role of fiber as a cancer-preventative factor in our diet.

■ FURTHER READING

Choi, S.-W. and Mason, J.B. (2002) Folate status: effects on pathways of colorectal carcinogenesis. *J. Nutr.* **132**:2413S–2418S.

Greenwald, P., Clifford, C.K. and Milner, J.A. (2001) Diet and cancer prevention. *Eur. J. Cancer* **37**:948–965.

Henderson, B.E. and Feigelson, H.S. (2000) Hormonal carcinogenesis. *Carcinogenesis* **21**:427–433.

Key, T.J., Allen, N.E., Spencer, E.A. and Travis, R.C. (2002) The effect of diet on risk of cancer. *Lancet* **360**:861–868.

Lamprecht, S.A. and Lipkin, M. (2003) Chemoprevention of colon cancer by calcium, vitamin D and folate: molecular mechanisms. *Nature Rev. Cancer* **3**:601–614.

Mathers, J.C. (2003) Nutrition and cancer prevention: diet–gene interactions. *Proc. Nutr. Soc.* **62**:605–610.

Nicholls, H. (2002) Aromatase inhibitors continue their ATAC on tamoxifen. *Trends Mol. Med.* **8**:S12–S13.

Qi, R. and Wang, Z. (2003) Pharmacological effects of garlic extract. *Trends Pharm. Sci.* **24**:62–63.

Venkitaraman, A.R. (2002) Cancer susceptibility and the functions of BRCA1 and BRCA2. *Cell* **108**:171–182.

■ WEB SITES

World Cancer Research Fund www.wcrf.org.uk

■ SELECTED SPECIAL TOPICS

Bingham, S.A., Day, N.E., Luben, R. and Ferrari, P. (2003) Dietary fibre in food and protection against colorectal cancer in the European Prospective Investigation into Cancer and Nutrition (EPIC): an observational study. *Lancet* **361**:1496–1501.

Brodie, A. (2002) Aromatase inhibitors in breast cancer. *Trends Endocrin. Met.* **13**:61–65.

Bub, A., Watzl, B., Blockhaus, M., Briviba, K.L, Liegibel, U., Muller, H., Pool-Zobel, B.L. and Rechkemmer, G. (2003) Fruit juice consumption modulates antioxidative status, immune status, and DNA damage. *J. Nutr. Biochem.* **14**:90–98.

Hites, R. A., Foran, J.A, Carpenter, D.O., Hamilton, M.C., Knuth, B.A. and Schwager, S.J. (2004) Global assessment of organic contaminants in farmed salmon. *Science* 303:226–229.

Naasani, I., Oh-hashi, F., Oh-hara, T., Feng, W.Y., Johnston, J., Chan, K. and Tsuruo, T. (2003) Blocking telomerase by dietary polyphenols is a major mechanism for limiting the growth of human cancer cells *in vitro* and *in vivo*. *Cancer Res.* 63:824–830.

Russo, J., Lareef, M.H., Balogh, G., Guo, S. and Russo, I.H. (2003) Estrogen and its metabolites are carcinogenic agents in human breast epithelial cells. *J. Steroid Biochem. Mol. Biol.* 87:1–25.

Ueland, P.M., Hustad, S., Schneede, J., Refsum, H. and Vollset, S.E. (2001) Biological and clinical implications of the MTHFR C677T polymorphism. *Trends Pharm. Sci.* **22**:195–201.

Wang, L.H., Yang, X.Y., Zhang, X., Mihalic, K., Fan, Y.-X., Xiao, W., Howard, O.M.E., Appella, E., Maynard, A.T. and Farrar, W.L. (2004) Suppression of breast cancer by chemical modulation of vulnerable zinc fingers in estrogen receptor. *Nature Med.* **10**:40–47.

Yue, W., Santen, R.J. Wang, J.-P., Li, Y., Verderame, M.F., Bocchinfuso, W.P., Korach, K.S., Devanesan, P., Todorovic, R., Rogan, E.G. and Cavalieri, E.L. (2003) Genotoxic metabolites of estradiol in breast: potential mechanism of estradiol induced carcinogenesis. *J. Steroid Biochem. Mol. Biol.* 86:477–486.

10

Cancer in the future: focus on diagnostics and immunotherapy

Introduction

This concluding chapter will address two issues: first is the question of whether cancer will exist in the future and, secondly, if the answer is "yes", what changes in cancer treatment and management are likely to be implemented? Overall, evidence suggests that cancer will "always be around" because mutation underlies carcinogenesis and we cannot escape from mutations. Although we may be able to avoid certain carcinogenic agents (e.g. tobacco) and processes (e.g. sunbathing), we certainly cannot avoid all of them.

Furthermore, cancer is associated with aging and life expectancies are increasing. As a consequence of people living longer, the incidence of cancer is increasing. On the other hand, maybe there is a lesson to be learnt from the history of medicine. In the past, there were dreaded infectious diseases such as smallpox, which are now preventable through vaccination. Could vaccination be used to eradicate cancer? The observation that the immune system could recognize and respond to tumors following bacterial infection was made over 100 years ago. More recent studies have shown that effector cells of the immune system can recognize tumor-associated antigens and kill tumor cells. This endogenous mechanism of protection against tumor cells by the immune system is called "immunosurveilance" and suggests that boosting the immune system by vaccination against tumor cells may be possible.

Alternatively, if we are not able to eradicate cancer, what will it be like having cancer in future decades. It is envisaged that cancer, a genetic disease at the cellular level, will be detected much earlier than is possible today because of the rise of genomics and its associated technologies. Instead of a positive diagnosis being linked imminently with death, cancer may become a long-term, chronic disease such as arthritis. Although a cure is preferred, the complexity of cancer may foster the development of treatments that allow people to live more comfortably with the disease rather than cure it.

In this chapter, we will examine current, far-reaching advancements in the fields of immunology and technology in order to form an educated prediction of the future of cancer. The potential for the prevention of cancer through vaccination will be examined. We will also investigate improved tools, techniques, and therapeutics that are essential in order to transform cancer into a chronic, rather than a terminal, disease. These include molecular diagnostics, expression profiling for cancer classification, bioinformatics, and new treatment strategies including cancer vaccines as therapeutic agents. Finally, towards the end of the chapter, a description of the people, agencies, and processes involved in cancer research is presented—just in case I have caught your interest in a career in cancer research.

10.1 Cancer vaccines

Our ability to harness the immune system to prevent and/or kill tumor cells is becoming evident. Vaccination is called active immunization because it tries to stimulate the individual's own immune effector cells. A vaccine is composed of antigen(s) and adjuvant(s). Adjuvants are vaccine additives that enhance the immune response to an antigen. This contrasts passive immunization, which involves the transfer of effectors of the immune system, such as T cells or secreted products of lymphoid cells, into the patient.

A little review of immunology basics

There are two main classes of lymphocytes (immune cells): Bone-marrow-derived (B) lymphocytes and thymus-derived (T) lymphocytes. The main function of B lymphocytes is to synthesize and secrete antibodies. Antibodies contain an antigen binding domain and overall, specific antibodies can recognize almost any antigen encountered. An antigen can be defined as any molecule that is able to generate an immune response. Antibodies can activate cell-mediated antibody dependent cell lysis. Since many B cells respond to an antigen, a mixture of antibodies is produced by many clones (polyclonal). Experimentally, we can grow a single clone of a specific B lymphocyte by creating a hybridoma in order to produce quantities of a specific, monoclonal, antibody. T cells are responsible for cell-mediated immunity. It is thought that the cytotoxic T cell response is the principal anti-tumor defense of the body. Cytokines are polypeptides involved in cell signaling in the immune system.

Passive immunization

Many early attempts at cancer immunotherapy utilized passive immunization strategies. Interferon-γ, a cytokine with promising ability to modulate the immune response, was studied extensively in phase I, II, and III trials. In general, however, poor clinical responses were observed. Tumor necrosis factor, also gave disappointing results in clinical trials. Recently, however, Rosenberg (see Box: Leaders in the Field) and his group have demonstrated promising results after the transfer of selected tumor-reactive T cells into cancer patients who underwent lymphodepleting chemotherapy (destruction of endogenous lymphocytes by cytotoxic drugs) (Dudley *et al.*, 2002). Tumor-infiltrating lymphocytes were expanded *in vitro* and transferred to patients along with the cytokine adjuvant, interleukin-2, for immunization. Cancer regression was observed in patients with metastatic melanoma.

Here, we focus on vaccinations. Cancer vaccines can either be designed to prepare the immune system prior to getting cancer for cancer prevention, so-called prophylactic vaccines, or they can act to stimulate the immune system in order to cause tumor regression in a patient with cancer, a therapeutic vaccine. Most cancer vaccines are designed to be therapeutic vaccines, though we will consider both types below.

Therapeutic vaccines

The production of a vaccine involves the selection of an appropriate antigen that will stimulate an effective anti-tumor response. Tumor-associated antigens may be derived from either degradation and processing of unfolded intracellular proteins that are shuttled to the surface of the tumor cell or from damaged or dying tumor cells. These may include oncoproteins arising from oncogenic mutations or chromosomal translocations. Since T cells are the main effectors of an anti-tumor response, antigens from the vaccine must be displayed eventually on the surface of other cells, called antigen-presenting cells. A series of cellular events characterize an immune response upon administration of a cancer vaccination (Figure 10.1). Antigen-presenting cells, such as dendritic cells that reside in the tissue, are at the heart of signaling for the mission of eliciting T-cell-mediated immunity. It is the dendritic cells that 1) acquire and 2) process the antigens and upon maturation migrate to the lymphoid organs to 3) present the antigens to the main effector T cells. The uptake of antigens by the dendritic cells is primarily by endocytosis. Antigen processing involves cleavage of the antigen into small peptides by proteases (depicted by scissors in Figure 10.1). The adjuvant in a cancer vaccine induces the maturation of the antigen-presenting cells and their migration to the lymphoid organs. Processed antigen is translocated to the cell surface for presentation in association with proteins

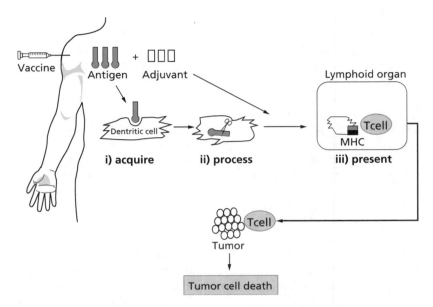

Figure 10.1 Cellular events of an immune responsé following vaccination

from the major histocompatibility complex (MHC; details of which are beyond the scope of this book). It is the CD8+ cytotoxic T cells that recognize the antigen on the tumor cell membrane and proceed to kill the tumor cells by releasing cytotoxic granules or inducing apoptosis.

Cancer vaccines are required to overcome tumor protective mechanisms. Tumor cells possess mechanisms, such as the secretion of particular growth factors, which allow them to evade and suppress the immune system. For example, transforming growth factor β obstructs dendritic cell maturation and interferes with antigen presentation to T cells. Let us examine several strategies used for producing cancer vaccines.

Whole-cell vaccines

Vaccines against infectious diseases are composed of bacteria or viruses whose ability to produce disease has been reduced or attenuated by different processes such as passage through an unnatural host, chemical treatment, or irradiation. The first cancer vaccines were composed of irradiated tumor cells, being modeled after successful, attenuated pathogen vaccines.

PAUSE AND THINK

The term vaccination comes from the Latin word "vacca" or cow because the first vaccine, reported in 1798 by E. Jenner, used cowpox virus for immunization against smallpox. As a doctor in the English countryside during a smallpox outbreak, Jenner noticed that milk maidens were less likely to contract smallpox, although these women were often exposed to cowpox infection. He hypothesized that cowpox infection was the cause of the resistance to smallpox and carried out experiments that supported his hypothesis.

All of the antigens expressed by a specific tumor are included in the whole-cell vaccine design. These first cancer vaccines demonstrated immune responses in mouse models but were disappointing in clinical trials causing either a weak response from the immune system (a weak immunogenic response) or a response against normal cells (autoimmunity). This may be due to the under-representation of immunogenic antigens relative to the total amount of antigens and stimulation against normal gene products, respectively. For example, vitiligo, an autoimmune disease that targets melanocytes, was observed in studies of a melanoma vaccine suggesting that the induced immune response also targeted normal antigens and thus normal cells. Some modifications of whole-cell vaccines are being pursued. For example, gene modified tumor cells that express stimulatory molecules for T cells double as antigens and adjuvants. However, regardless of their degree of success, whole-cell vaccines have been important stepping-stones towards antigen-specific vaccines.

Peptide-based vaccines

Another strategy for cancer vaccine development is to use tumor-associated antigens to generate an immune response. This involves the identification and characterization of specific molecules on the tumor cells that are recognized by T cells rather than using whole cells from tumors as was described above. Tumor-specific antigen molecules have qualitative or quantitative differential expression patterns in tumor cells compared with normal cells. Many of these antigens elicit an immunogenic response without autoimmunity. This has led to the production of antigen-specific peptide vaccinations. The peptides used are short sequences of amino acids that code for a part of the tumor-associated antigen and can be produced as synthetic or recombinant proteins.

A growing list of breast tumor antigens, including HER2, mucin1, and carcinoembryonic antigen (CEA), provide the basis for the production of breast cancer vaccines. Several melanoma tumor antigens have also been characterized. A peptide-based vaccine targeting the melanoma-associated antigen glycoprotein 100 (gp100) was successful in producing a therapeutic clinical response in melanoma patients as reported by Rosenberg (see Box below) and colleagues. The gp100 antigen is an antigen that is expressed in normal melanocytes, melanomas, and pigmented retinal cells. The study used a modified gp100 peptide that had an increased ability to generate reactive cytotoxic T cells, along with cytokine

PAUSE AND THINK

What is the difference in methods between making synthetic and recombinant protein? Synthetic peptides are made in the laboratory by linking amino acids together in a specific order while recombinant proteins are synthesized *in vivo* from genetically engineered molecules that include a DNA sequence encoding the specified amino acids.

adjuvant, IL-2. Forty-two percent of patients (13 of 31 people) demonstrated cancer regression in metastases from different locations.

A leader in the field of vaccination: Steven A. Rosenberg

The Institute for Scientific Information reported in 1999 that Rosenberg was the most cited clinician in the world in the field of oncology for the 17 years between 1981 and 1998. Rosenberg is the author of over 820 scientific articles covering various aspects of cancer research and has authored eight books.

Rosenberg helped to develop the first effective immunotherapies for selected patients with advanced cancer. He was also the first person successfully to insert foreign genes into humans, pioneering the development of gene therapy for the treatment of cancer. Along with his research group, he cloned the genes encoding cancer antigens and used these as the basis to develop cancer vaccines for the treatment of patients with metastatic melanoma. His recent studies, involving the transfer of anti-tumor lymphocytes and their repopulation in cancer patients, demonstrated cancer regression.

Rosenberg received his B.A. and M.D. degrees at Johns Hopkins University in Baltimore, Maryland and a Ph.D. in Biophysics at Harvard University. After completing his residency training in surgery in 1974, Dr Rosenberg became the Chief of Surgery at the National Cancer Institute, a position he still holds at the present time. Rosenberg is also currently a Professor of Surgery at the Uniformed Services University of Health Sciences and at the George Washington University School of Medicine and Health Sciences in Washington, D.C.

Dendritic cell vaccines

Vaccines may also be composed of human dendritic cells, cells that are critical antigen-presenting and stimulatory cells for the induction of a T-cell dependent immune response. Dendritic cells originate in the bone marrow, and reside in an immature state in peripheral tissues. As described above, upon receiving inflammatory signals, they differentiate or mature and migrate to lymph nodes where antigens are presented and the T-cell response is initiated. *In vivo*, tumors secrete several factors that suppress dendritic cell differentiation and migration, and may contribute to the immunosuppression observed in cancer patients.

For the purpose of vaccination, dendritic cells must be isolated from an individual patient and cultured *in vitro* during which time they can be loaded or pulsed with specific antigens, DNA, or RNA via their high capacity for endocytosis (Figure 10.2) (or other means of transfection such as electroporation). Subsequently, they are reintroduced back into the patient. Thus, dendritic vaccines are labor intensive and expensive. Initial clinical trials using loaded dendritic cells have shown positive clinical responses and no significant toxicity. An antigen-loaded dendritic

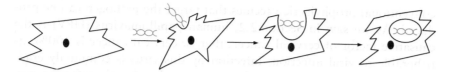

Figure 10.2 Dendritic cell loading

cell vaccine called Provenge (Dendreon Corporation, Seattle, WA) was produced for the treatment of prostate cancer by the following steps:

(i)   a dendritic cell precursor-enriched fraction was isolated;

(ii) the cells were matured *in vitro* by incubation with a recombinant fusion protein (consisting of prostatic acid phosphatase linked to granulocyte-macrophage colony-stimulating factor (GM-CSF)) that targets the GM-CSF receptor present on dendritic cells;

(iii) the mature dendritic cells, now carrying the prostate cancer antigen, is administered.

Phase III trial results demonstrated an effect on time to disease progression in some patients with intermediate disease (http://www.dendreon.com). Dendritic cell vaccines continue to be actively investigated. CEA (mentioned above) is also overexpressed in more than 95% of colorectal cancers and has provided a target for colorectal cancer vaccination. CEA RNA pulsed dendritic cell vaccines are in clinical trials. Ongoing clinical trials for colorectal cancer can be viewed at the National Cancer Institutes web site (www.nci.nih.gov). Expanding and loading dendritic cells *in vivo*, though not yet possible, is a promising idea for the future that would eliminate the dangers (e.g. contamination) associated with cell transfers and reduce labor and costs.

Overall the optimization of therapeutic vaccines needs to be pursued. The correct patient population (i.e. with respect to age, cancer stage, molecular signature of tumors) and timing of vaccine administration needs to be a focus. These vaccines may also become a tool for cancer management since many trials have shown vaccination results more often in a stable disease response rather than a complete curative response. Side effects of vaccines are usually minimal in contrast to conventional chemotherapies.

Vaccines for cancer prevention

The therapeutic vaccines discussed above are aimed at the tumor. Vaccines generated from shared tumor antigens have been successful as prophylactic vaccines in animal models but have not been tested in humans. However, there are a few select types of cancer that are caused by pathogenic carcinogens (i.e. bacteria or viruses) and in these cases, conventional prophylactic vaccines that target the pathogen can be produced. As we saw in Section 2.2, human papillomavirus (HPV) is the causative factor of cervical cancer; that is, cervical cancer is 100% attributable to viral infection. Infection by this virus is commonly eradicated by the immune system, but more rarely, chronic infection persists and leads to cervical cancer. There are several high-risk virus types, with HPV16 accounting for more than 50% of cervical cancers. Several vaccinations against HPV16, have been produced and have given promising results in clinical trials. The major viral capsid protein, L1, after expression in eukaryotic cells, self-assembles into virus-like particles and elicits an immune response. These vaccines prevent persistent infection by the virus, and thereby potentially prevent cervical cancer. This translates into preventing 150,000 deaths per year in developing countries alone.

Large strides are being made in the development of prophylactic vaccines for breast cancer. As mentioned above, several promising breast-cancer antigens have been characterized. Prophylactic breast cancer vaccines are likely to be an important alternative to prophylactic mastectomies and/or oophorectomies or chemoprevention in women who carry germline mutations in the BRCA1 and BRCA2 genes (see Section 9.5). Safety and immune responses have been demonstrated in several Phase I and II trials but large trials are needed. Reluctance to carry out large-scale trials comes from the fear of autoimmunity against normal breast tissue, though autoimmune attack of normal breast tissue may be tolerable and may not have more severe consequences than mastectomies.

Hurdles to jump

There are several problems that need to be overcome for the full potential of vaccine development to be reached. First, the immune system becomes less effective with aging and is suppressed by conventional chemotherapy. It is rare that preclinical studies are performed in old mice or mice who were pretreated with chemotherapy and this may help to explain the discrepancies between outcomes in mice and man; positive immunological responses in mice are often not reproducible in humans. It may be that therapeutic cancer vaccines may be more successful in pediatric cancer patients compared with older patients. Such comparisons need to be carried out. Secondly, many vaccines may be most effective in early stage cancer patients, although trials using such patients are unlikely to receive approval. In addition, resistance against therapeutic vaccines may arise. Antigen-negative tumor cell clones evolve due to selective pressure exerted by the vaccine. Mutations that alter antigen expression will allow tumor cells to evade the immune response and survive.

Also, vaccines need to be tested in all appropriate contexts. Vaccines against tumor-specific antigens are being tested in humans exclusively as therapeutic agents, and not as prophylactics, even though the success of these agents in preclinical trials has been demonstrated almost exclusively as prophylactics. Note that since human tumors can only be grown in immune-deprived mice (e.g. nude mice), immunotherapy studies on human tumors cannot be performed in existing preclinical models and results from animal models may be species-specific. We cannot assume that what is successful in mice will be successful in man because some aspects of physiology between the two are different. Prophylactic vaccines aimed at tumor specific antigens (not including those directed against pathogens, e.g. HPV) have not been tested in clinical trials because the test population will be healthy individuals and the consequences and/or side effects are unknown. However, at some point, vaccines as prophylactics need to be tested in humans.

10.2 Microarrays and expression profiling

Many of the therapies described in this book will only be successful on tumors that have the appropriate molecular profile. For example, Herceptin and Erbitux will be effective for tumors that overexpress HER2 and BCL2 antagonists such as Genasense will be effective for tumors that overexpress BCL2. A diagnostic to identify EGFR-positive tumors (Dako; Copenhagen) has received approval for use with Erbitux. The more molecular information that can be obtained from a tumor, the more precise the treatment that can be administered. Thus, it is essential to develop quick and inexpensive methods for analyzing the molecular profiles of tumors.

Microarrays and their associated technologies have enabled the expression of thousands of genes to be analyzed easily. This technique is a tremendous asset providing data sets that are not available by other methods. Previous methods could only examine individual genes or at most, small sets of genes at a time. The applications of microarrays in cancer biology are far-reaching. Since cancer is a genetic disease at the cellular level, the changes in gene expression of genes that are involved in carcinogenesis can be identified. Different gene expression signatures can be identified for specific cancers. The data generated may suggest new tumor classifications and new molecular targets for the development of cancer therapeutics. The molecular signatures may also allow for the prediction of disease outcome and prescription of the most efficient treatment available for a particular tumor type. A futuristic vision is to be able to develop tailor-made therapies for individual patients based on the genetic profile of their primary tumor.

Experimental procedure

Microarrays are grids, usually made on glass slides or silicon chips. They hold DNA representing thousands of genes that act as probes (i.e. sequences that are complementary and can hybridize to specific RNAs) for RNA. The RNAs that are in a sample indicate genes that are expressed (transcribed) in that sample. A typical protocol will be described. Thousands of gene-specific hybridization probes are applied to a glass slide or silicon chip (Figure 10.3a). The DNA is usually bound to defined locations on the grid by robotic or laser technology. RNA is isolated from a biological sample, such as a tumor and copied to incorporate fluorescent nucleotides or a fluorescent tag (Figure 10.3b). The chip is then incubated with labeled RNA or complementary DNA (cDNA) from the tumor sample (Figure 10.3c). Unhybridized RNA is washed off and the microarray is then scanned under a laser and analyzed by computer (Figure 10.3d). A sample microarray image is shown in Figure 10.3e. By analyzing the fluorescent intensities of the RNA or cDNA hybridized to the probes using computerized scanners, gene expression can be quantified. There are two common types of microarrays: cDNA microarrays

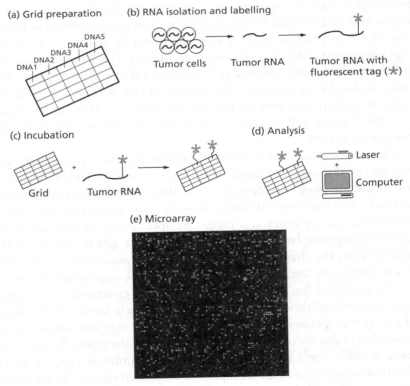

Figure 10.3 (a–d) The basic protocol for microarrays; (e) Sample microarray

and oligonucleotide microarrays. The difference between the two is due to the nature of the probes. In cDNA microarrays each probe has its own ideal hybridization temperature (based on factors such as GC content) and thus intensities of a test sample must always be compared with a control sample processed at the same time. In oligonucleotide microarrays, the synthetic probes are designed such that all the probes have identical hybridization temperatures allowing absolute values of expression to be measured within one sample. Results from microarrays can be visualized in different formats. One type of data display format is a heat map that uses color to represent the levels of gene expression. Modifications of the technique, including protein microarrays and antibody microarrays, demonstrate that the microarray design has far-reaching potential.

Application of microarrays to cancer classification

Let us examine one example of how a microarray was applied to characterize subclasses of a specific cancer. The application of a lymphochip, a microarray that screens for genes important to cancer, immunology, and lymphoid cells, helped to define two molecularly distinct forms of a particular lymphoma (Alizadeh *et al.*, 2000). Diffuse large B-cell lymphoma is clinically heterogeneous whereby 40% of patients respond to treatment and 60% do not. Results from the lymphochip showed that there were two distinct patterns of gene expression that were differentiation-stage specific and also corresponded to different clinical outcomes (76% of one subgroup was alive after 5 years, compared with only 16% of the other). The classification of one cancer was refined to two distinct cancers. These findings have important implications for cancer management and stimulate the rethinking of the current treatment strategy. Currently all patients of diffuse large B-cell lymphoma undergo chemotherapy first. Those who have a poor response are then named candidates for bone marrow transplantation. Perhaps some patients, as defined by their molecular signature, should be directed for bone marrow transplantation immediately. DNA arrays are being used to refine molecular classifications of other cancers, including breast cancer.

10.3 Molecular diagnostics and prognostics

It is well established that early diagnosis of cancer is crucial for a good prognosis. Vigorous cytology screening programs (cervical smear tests) in developed countries have resulted in markedly reduced cervical cancer death rates. However, the power of molecular diagnostics is only beginning to be applied and promises to make major contributions to the field. Microarrays will play a prominent role in diagnosis. Interestingly, approximately 5% of gene expression distinguishes any normal tissue from its corresponding cancerous tissue. In addition, precancerous lesions can be distinguished from normal and cancerous tissue by expression profiling. These facts can be applied to diagnostics and prognostics.

Let us examine some progress that has been made for the detection of prostate cancer, the second leading cause of cancer-related deaths in men. Prostate-specific antigen (PSA) has been the conventional prostate tumor marker whereby elevated levels can be detected in blood. However, improvements are needed to prevent subsequent negative biopsy rates (70–80%). Negative biopsy rates are due to the fact that PSA is not specific for prostate cancer. Elevated levels are also detected in benign prostate conditions such as prostatitis. Diagnostics based on genomics are leading to improvements in prostate cancer detection. DD3^{PCA3} is a gene that has been identified as the most prostate cancer-specific gene described thus far. DD3^{PCA3} is only expressed in prostate tissue and is strongly over-expressed (10–100-fold) in greater than 95% of prostate tumors. Urinalysis that detects DD3^{PCA3} RNA, using a nucleotide amplification method called quantitative reverse transcriptase-PCR, has been designed for the diagnosis of prostate cancer. The test has been carried out on urine samples collected after prostate massage to help transport tumor cells into the urethra and the results (67% sensitivity) hold great promise as a non-invasive diagnostic tool that may reduce the number of unnecessary biopsies (Hessels *et al.*, 2003). Cancer blood tests that can provide comprehensive molecular information may one day become a reality with further study focused around the observation that circulating DNA from tumor cell death is released into the bloodstream. Another futuristic idea is that it may become possible to implant a gene chip under the skin to monitor changes in the expression of specific genes, thus speeding up diagnosis and facilitating early treatment.

As we saw in Chapter 3, hypermethylation of particular gene promoters is characteristic of specific tumors. The promoter regions of two genes, p16 and O^6-methyl-guanine-DNA methyltransferase, are frequently methylated in lung cancer. The p16 gene product is a tumor suppressor that plays a role in the regulation of the cell cycle while the latter is involved in the repair of DNA damage caused by alkylating agents. Methylation-specific PCR (see Chapter 3) has been used to detect the aberrant methylation of these two genes in the sputum of patients both at the time of lung cancer diagnosis and up to three years prior to diagnosis (Palmisano *et al.*, 2000). This suggests that aberrant DNA methylation may be an applicable non-invasive molecular diagnostic marker.

Several groups have used microarrays to identify a number of marker genes whose expression can predict metastasis and/or prognosis (Shipp *et al.*, 2002; van't Veer *et al.*, 2002). As a result, the first patient gene expression profiling tests were launched in 2004. Oncotype DX (Genomic Health, Redwood City, CA) and Mammaprint (Agendia, Amsterdam) have designed tests that can predict breast cancer progression. Sixteen genes selected from a total of 250 and 70 genes from a total of 25,000, respectively, have been identified to indicate good or bad prognosis. The main purpose of analyzing these selected genes is to identify tumors that are unlikely to metastasize in order to spare patients the trauma of chemotherapy. Only a small proportion of patients whose breast cancer has not spread to the lymph nodes develop metastases (20%) and really require chemotherapy. Most node-negative breast cancer patients are cured by surgery and radiotherapy. Until a gene expression profiling test is generally in place, and its ability to vigorously distinguish between the two groups in the clinic confirmed, many patients will be receiving unnecessary chemotherapy in order to prevent metastasis in the unidentified few, for whom it is essential.

Mutational analysis can help predict drug sensitivity

The gene sequence of a molecular target for a particular drug can help predict drug sensitivity. It has been demonstrated that sensitivity to Iressa (Gefitinib; AstraZeneca) depends on mutations in the *EGFR* (EGF receptor) gene that codes for its molecular target. Iressa, a tyrosine kinase inhibitor targeted against the EGF receptor (see Section 4.3), was reported to have mixed clinical responses. Approved in the USA and Japan for the treatment of non-small cell lung cancer, Iressa causes impressive and successful tumor regression but, curiously, in only a select 10% of patients. Since non-small cell lung cancer accounts for 80% of all lung cancers, the number of patients is significant. It has now been demonstrated that the patients who respond to Iressa carry mutations in the *EGFR* gene (Paez *et al.*, 2004; Lynch *et al.*, 2004).

Clinical trials showed that the Japanese are three times more likely to respond to Iressa compared with Americans. Scientists screened for mutations in the *EGFR* gene in two groups of non-small cell lung cancer tumor samples: one set from a Japanese hospital and another set from an American hospital. Somatic mutations in the *EGFR* gene were found more often in Japanese patients. The mutations were either missense mutations or small deletions that were located in the coding region for the receptor tyrosine kinase active site, a region near the binding site for Iressa. Scientists then asked the question of whether the mutations correlated with treatment response. The analysis of a small set of patients showed that those who responded to Iressa also carried the identified mutations.

The mutations produce an altered protein that increases the amount and duration of EGF-induced signal transduction compared with wild-type receptors. In other words, the receptors are hyperactive but not constitutive. These mutations that underlie aberrant tyrosine kinase signaling and drive carcinogenesis also make the tumor more susceptible to Iressa because they cluster in the sequences that code for the region where Iressa binds. This suggests that knowing particular molecular characteristics of tumors (e.g. mutations of the *EGFR* gene) is important for selecting the best treatment for an individual. See Pause and Think.

Similarly, mutations in the *KIT* gene affect the sensitivity of gastrointestinal stromal tumor (GIST) to Gleevec (Glivec). These and similar observations support the emergence of pharmacogenetics, the study of correlating specific gene DNA sequence information to drug response.

PAUSE AND THINK

Why are these particular mutations in the *EGFR* gene more common in Japanese than Americans? It may be due to a lifestyle factor such as exposure to a specific carcinogen or a genetic predisposition, or a mixture of both.

10.4 Cancer research bioinformatics

The time will soon come when hospitals and health services will be easily able to subject an individual tumor to genomic analysis. Such analysis could take place over the course of the tumor's known history: upon detection, during and after treatment (Figure 10.4). This type of information could be pooled from thousands of geographic locations and thousands of clinical trials, providing an extraordinary tool for future treatment and research. The molecular profile of an individual patient will be able to be compared and analyzed in order to select the best-known therapy available. Researchers will be able to identify cancer-specific molecular targets for drug design more rapidly. Several new bioinformatics initiatives have been launched in 2004. The US National Cancer Institute and the UK National Cancer Institute will collaborate in developing tumor information databases and networks (see web sites below). These initiatives will make tissue data, biological contexts or ontologies, and clinical trial information readily available to researchers. Computer capabilities and facilities will need to be expanded and enhanced to handle the data generated. Some improvements in image compression have already been made allowing histology images to be analyzed over the internet.

10.5 Improved clinical trial design

There are many flaws in the clinical trials performed today that result in an inflated number of unsuccessful drug results. The most difficult problem to overcome is that many new drugs are tested on older patients that

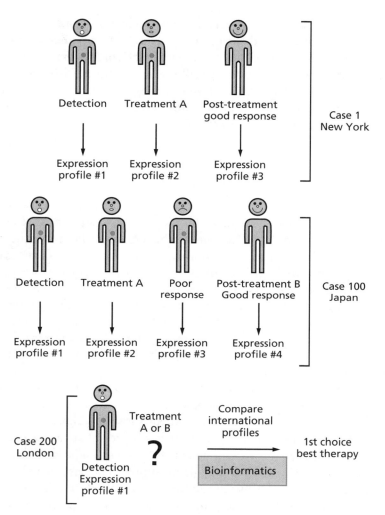

Figure 10.4 Cancer bioinformatics: a possible application for administration of first choice best therapy

have advanced cancer and who have not responded to a range of conventional therapies. Ethically acceptable suggestions are needed on how we can move forward.

The results above that illustrated variable response rates among two different populations to the drug Iressa, argue for the benefit of including geographically and genetically diverse populations in clinical trials. Molecular profiling of the tumor prior to, and/or during, clinical trials is also important in order to gain a true measure of drug efficacy. Patient selection (i.e. choosing patients that have the molecule or molecular defect that a drug is designed to target) is necessary so that drugs are shown to be efficient when tested on the "right" patients. For example, as we saw above with Iressa, if a drug acts on tumor cells that contain a mutated receptor, the efficacy of the drug should be tested on patients whose tumor contains these mutations. Otherwise a variable response will be observed and the observed efficacy will not be accurate. Also, some studies have shown that expression levels of drug targets can change during progression of the disease and therefore drug response may be expected to change during the course of the disease.

10.6 Treating cancer symptoms

In order for cancer to become a chronic disease, symptoms that are associated with it and that contribute to the high mortality rates and poor physiological function must be addressed. The alleviation of cancer pain is one area towards which many pharmaceutical companies are directing their research. Successful drugs will greatly increase the quality of life of cancer patients. Cachexia, a metabolic defect that is characterized by progressive weight loss due to the deletion of adipose tissue and skeletal muscle, is another common symptom of cancer. It is the early loss of skeletal muscle that makes cachexia dissimilar to starvation. In starvation, fat is lost prior to muscle in order to preserve lean body mass. The mechanism for cachexia involves the upregulation of catabolism or a defect in anabolism. Nutritional supplements and inducers of appetite are unable to relieve the process. Since patient survival is directly related to the total weight loss, therapeutic strategies based on anabolic and anti-catabolic pathways must be designed. Eicosapentaenoic acid (EPA) or fish oil combined with an energy-dense supplement has been shown to be clinically effective for increasing lean body mass. Fish oils repress the transcription of genes whose products play a role in protein catabolic pathways, such as the ubiquitin-proteasome pathway (Tisdale, 2002). Further work in this area is needed.

10.7 A career in cancer research: the people, agencies, and process involved in drug development

There are several avenues to pursue for a career in cancer research. The road most often traveled is to obtain a Ph.D. in an area that interests you. Interest is important for several reasons: first, you will be spending many hours reading and thinking about your subject; secondly, you will spend many hours in the laboratory conducting experiments which focus around your topic; thirdly, one can never predict in what area a big breakthrough will occur, so there is no sense in trying to guess. Research can be frustrating at times because you are trying to figure out something that no one knows or has really done before and progress tends to be in small steps. However, many small steps made by many individuals drives the field. It can be most rewarding to know that you have made contributions to relieving suffering and saving lives. Studentships are posted in the back of the leading scientific journals such as *Nature* and *Science*, but a personal approach by letter to the head of a laboratory in which you are interested in working can also be successful. Alternatively, cancer research can be pursued in a pharmaceutical company with entry levels at different stages of education.

Cancer research is expensive and requires appropriate facilities equipped with high-tech equipment. These facilities are available in universities, research institutes, hospitals, and in biotechnology and pharmaceutical companies (see Appendix 2). The relationship between these types of research providers has recently grown to be symbiotic; a close association leads to mutual benefit. Several organizations, such as the London Technology Network, create opportunities that help foster collaborations between universities and industry. Such organizations also provide commercial business training for academics. Several specialized publications (e.g. *Gibson Index Newsletter*, www.Gibson-Index.com) cover the news of small and medium enterprises and university spin out companies in the UK.

Drug development follows a series of stages (see Box below) and different stages can be performed by people of different expertise (e.g. biochemists, cell biologists, chemists, clinicians) at different facilities.

Stages of drug development

Molecular target → screen for inhibitors or activators → Optimization and Formulation → Preclinical and Clinical Trials → regulatory approval

---

Let us look at the development of Imatinib (Gleevec, USA; Glivec, UK, Europe) in relation to the stages of drug development (Capdeville *et al.*, 2002). A chromosomal translocation, characteristic of chronic myelogenous leukemia results in elevated tyrosine kinase activity that is crucial for transformation. The identification of the molecular consequence of this translocation pointed to the resulting fusion protein, BCR-ABL, as a molecular target. A lead compound (a compound that shows a desired activity, e.g kinase inhibition) was elucidated from a chemical screen that inhibited protein kinase C. Optimization of the lead compound by the addition of small chemical groups produced a drug that inhibited BCR-ABL tyrosine kinase activity and showed good bioavailability.

Bioavailability relates to the ability of the drug to reach its site of action after administration. Preclinical testing demonstrated fairly selective inhibition of BCR-ABL kinase activity and induction of apoptosis in cell culture experiments and in leukemic cells from patients. Inhibition of tumor growth was also observed in animal models. Clinical trials (see Table 1.1 for phases of clinical trials) documented safety and efficacy. Approval by the Food and Drug Administration (FDA) was given on May 2001, and subsequent approval for use in Europe and Japan followed shortly afterwards. The European Medicines Evaluation Agency (EMEA) and the Japanese Ministry of Health and Welfare, together with the FDA are working towards an internationally harmonized system for drug approval that will facilitate the delivery of drugs to more patients, more quickly.

10.8 Are we making progress?

Do you think we are making progress? Despite several media articles that raise doubts, the real answer to this question is almost certainly: Yes! The statistics are available. For example, the overall survival for all stages of prostate cancer combined has increased from 67% to 89% over the past 20 years. The increased survival is attributable to both earlier and better detection and advances in therapeutics. Although our knowledge about cancer has grown expansively, there is still so much more to learn. Perhaps there are some secrets held in the heart—literally. Primary cardiac tumors, of which only one quarter are malignant, are rare (0.02%). Investigations into why cancer is rare in this particular tissue may lead to knowledge of protective mechanisms that can be applied to other tissues.

You will notice that most of the newly approved therapies, shown in the list below, are directed against molecules that are tyrosine kinases (e.g. EGFR, VEGFR, ABL). There are several tyrosine kinases that are known to play important roles in carcinogenesis (e.g. fibroblast growth factor receptor, FGFR) but have yet to become targets for new drugs in clinical trials. However, although we can design new cancer therapies against molecular targets, tumor cells may undergo additional mutations that can result in drug-resistant clones. This suggests that combinations of drugs and drug strategies are important for future treatment regimens. As we saw in previous chapters, there are many potential molecular therapies, such as angiogenesis inhibitors, anti-endocrine drugs, apoptotic inducers, cell cycle inhibitors, HDAC inhibitors, and inhibitors of cell renewal signaling pathways, in development. [I regret that some strategies have not been discussed (e.g. proteasome inhibitors).]

For many of these drugs, the therapeutic index is enhanced compared with conventional chemotherapies We await the elongation of the list of newly approved molecular cancer therapeutics shown below which I predict will be sooner rather than later.

Targeted cancer therapeutics approved in 2004

Trademark	Drug	Description	Target	Cancer	Company
Avastin	Bevacizumab	humanized mAb	VEGF	colorectal	Genentech
Erbitux	Cetuximab	humanized mAb	EGFR	colorectal	Imclone
Iressa	Gefitinib	small mol.inhibitor	EGFR	NSCLC	Astra Zeneca
Velcade	Bortezomib	proteasome inhibitor		myeloma	Millennium Pharm.
Gleevec (USA) Glivec (UK, Europe)	Imatinib	small mol. inhibitor	BCR-ABL KIT, PDGFR	CML; GIST	Novartis
Herceptin	Trastuzumab	humanized mAb	HER2	breast	Genentech

CHAPTER HIGHLIGHTS—REFRESH YOUR MEMORY

- Cancer vaccines can be designed to be prophylactic or therapeutic.
- Therapeutic vaccines use whole-cells, peptides, or dendritic cell strategies.
- Conventional prophylactic vaccines can be aimed at cancer caused by pathogens, e.g. the human papillomavirus.
- Microarrays analyze the expression of thousands of genes at one time.
- Microarrays have several applications including identifying new oncogenes, helping to refine cancer classifications, and predicting cancer prognosis.
- Molecular profiling of individual tumors may allow for tailor-made therapies.
- Mutations within a molecular drug target can affect drug sensitivity.
- The response to Iressa is dependent on specific mutations in the *EGFR* gene.
- New bioinformatic initiatives have been launched to coordinate the organization and sharing of data from cancer studies.
- Molecular profiling is a tool that should be used in clinical trials in order to identify particular molecular targets in a tumor and test a drug's true efficacy.
- Cachexia is a metabolic defect resulting in the progressive loss of fat and protein, may be a symptom associated with cancer.
- Cancer research requires the talent and expertise of many people.
- We are making progress in the field of molecular cancer therapeutics.

■ ACTIVITY

An enzyme involved in the functional activity of phosphorylation sites, called Pin1 has been identified. Discuss the molecular mechanisms of Pin1 action, its role in oncogenesis, and its use as a molecular diagnostic tool. [Hint: Start with Lu, K.P. (2003). *Cancer Cell* **4**:175.]

■ FURTHER READING

Antonia, S., Mule, J.J. and Weber, J.S. (2004) Current developments of immunotherapy in the clinic. *Curr. Opin. Immun.* **16**:130–136.

Blattman, J.N. and Greenberg, P.D. (2004) Cancer immunotherapy: A treatment for the masses. *Science* **305**:200–205.

Figdor, C.G., de Vries, I.J., Lesterhuis, W.J. and Melief, C.J.M. (2004) Dendritic cell immunotherapy: mapping the way. *Nature Med.* **10**:475–480.

Finn, O.J. (2003) Cancer vaccines: Between the idea and the reality. *Nature Rev. Immunol.* **3**:630–641.

Finn, O.J. and Forni, G. (2002) Prophylactic cancer vaccines. *Curr. Opin. Immun.* **14**:172–177.

Frazer, I.H. (2004). Prevention of cervical cancer through papillomavirus vaccination. *Nature Rev. Immunol.* **4**:46–54.

Liu, E.T. (2003) Classification of cancers by expression profiling. *Curr. Opin. Gen. Dev.* **13**:97–103.

Schuler, G., Schuler-Thurner, B. and Steinman, R.M. (2003) The use of dendritic cells in cancer immunotherapy. *Curr. Opin. Immun.* **15**:138–147.

Zeh III, H.J., Stavely-O'Carroll, K. and Choti, M.A. (2001) Vaccines for colorectal cancer. *Trends Mol. Med.* **7**:307–313.

■ WEB SITES

Bioinformatics initiatives

Cancer Biomedical Informatics Grid http://cabig.nci.nih.gov and

www.cancerinformatics.org.uk

Clinical trials

National Cancer Institute www.nci.nih.gov

■ SELECTED SPECIAL TOPICS

Alizadeh, A.A., Eisen, M.B., Davis, R.E., Ma, C., Lossos, I.S., Rosenwald, A., Boldrick, J.C., Sabet, H., Tran, T., Yu, X., Powell, J.I., Yang, L., Marti, G.E., Moore, T., Hudson Jr., J., Lu, L., Lewis, D.B., Tibshirani, R., Sherlock, G., Chan, W.C., Greiner, T.C., Weisenburger, D.D., Armitage, J.O., Warnke, R., Levy, R., Wilson, W., Grever, M.R., Byrd, J.C., Botstein, D., Brown, P.O. and Staudt, L.M. (2000) Distinct types of diffuse large B-cell lymphoma identified by gene expression profiling. *Nature* **403**:503–511.

Capdeville, R., Buchdunger, E., Zimmermann, J. and Matter, A. (2002) Glivec (STI571, Imatinib), a rationally developed, targeted anticancer drug. *Nature Rev. Drug Discovery* 1:493–502.

Dudley, M.E., Wunderlich, J.R., Robbins, P.F., Yang, J.C., Hwu, P., Schwartzentruber, D.J., Topalian, S.L., Sherry, R., Restifo, N.P., Hubicki, A.M., Robinson, M.R., Raffeld, M., Duray, P., Seipp, C.A., Rogers-Freezer, L., Morton, K.E., Mavroukakis, S.A., White, D.E. and Rosenberg, S.A. (2002) Cancer regression and autoimmunity in patients after clonal repopulation with anti-tumor lymphocytes. *Science* 298:850–854.

Hessels, D., Klein Gunnewiek, J.M.T., van Oort, I., Karthaus, H.F.M., van Leenders, G.J.L., van Balken, B., Kiemeney, L.A., Witjes, and J.A. and Schalken, J.A. (2003) DD3^{PCA3}—based molecular urine analysis for the diagnosis of prostate cancer. *Eur. Urol.* 44:8–16.

Lynch T.J., Bell, D.W., Sordella, R., Gurubhagavatula, S., Okimoto, R.A., Brannigan, B.W., Harris, P.L., Haserlat, S.M., Supko, J.G., Huluska, F.G., Louis, D.N., Christiani, D.C., Settleman, J. and Haber, D.A. (2004) Activating mutations in the epidermal growth factor receptor underlying responsiveness of non-small cell lung cancer to gefitinib. *N. Engl. J. Med.* 350:2129–2139.

Paez, J.G., Ja:nne, P.A., Lee, J.C., Tracy, S., Greulich, H., Gabriel, S., Herman, P., Kaye, F.J., Lindeman, N., Boggon, T.J., Naoki, K., Sasaki, H., Fujii, Y., Eck, M.J., Sellers, W.R., Johnson, B.E. and Meyerson, M. (2004) EGFR mutations in lung cancer: correlation with clinical response to Gefitinib Therapy. *Science* 304:1497–1500.

Palmisano, W.A., Divine, K.K., Saccomanno, G., Gilliland, F.D., Baylin, S.B., Herman, J.G. and Belinsky, S.A. (2000) Predicting lung cancer by detecting aberrant promoter methylation in sputum. *Cancer Res.* 60:5954–5958.

Rosenberg, S.A. (2001) Progress in human tumor immunology and immunotherapy. *Nature* 411:380–384.

Shipp. M.A., Ross, K.N., Tamayo, P., Weng, A.P., Kutok, J.L., Aguiar, R.C.T., Gaasenbeek, M., Angelo, M., Reich, M., Pinkus, G.S., Ray, T.S., Koval, M.A., Last, K.W., Norton, A., Lister, A., Mesirov, J., Neuberg, D.S., Lander, E.S., Aster, J.C. and Golub, T.R. (2002) Diffuse large B-cell lymphoma outcome prediction by gene-expression profiling and supervised machine learning. *Nature Med.* 8:68–74.

Tisdale, M.J. (2002) Cachexia in cancer patients. *Nature Rev. Cancer* 2:862–870.

Van't Veer, L.J., Dai, H., van de Vijver, M.J., He, Y.D., Hart, A.A.M., Mao, M., Peterse, H.L., van der Kooy, K., Marton, M.J., Witteveen, A.T., Schreiber, G.J., Kerkhoven, R.M., Roberts, C., Linsley, P.S., Bernards, R. and Friend, S.H. (2002) Gene expression profiling predicts clinical outcome of breast cancer. *Nature* 415:530–535.

APPENDIX 1: CELL CYCLE REGULATION

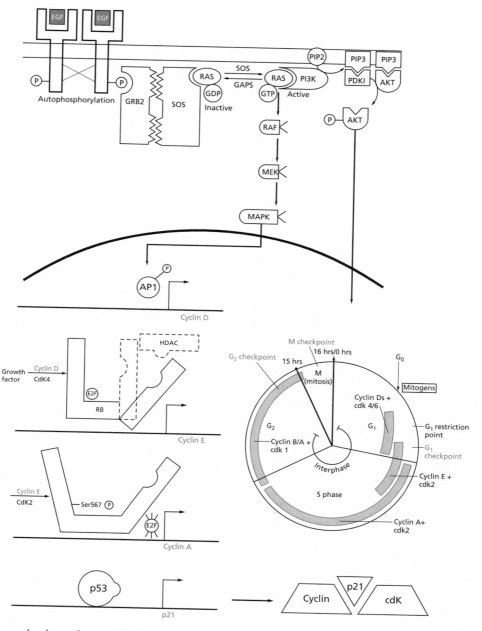

Appendix 1 Key molecular pathways of cell cycle regulation. Growth factor signaling results in the expression of target genes, including cyclin genes. Cyclin proteins are key for progression of the cell cycle and are involved in the regulation of the tumor suppressor protein, Rb. The protein product of the *p21* gene acts as an inhibitor of cyclin-cdk (cyclin-dependent kinase) complexes. These pathways are discussed in detail in the text.

■ APPENDIX 2: CENTERS FOR CANCER RESEARCH

USA

Cancer Research Laboratory,
University of California, Berkeley
447 Life Sciences Addition MC2751
94720-2751 Berkeley
California, USA
Tel.: 1510 642 4711
Fax: 1510 642 5741
Central Email: cslatten@uclink4.berkeley.edu
Central URL: http://biology.berkeley.edu/crl/

Cold Spring Harbor Laboratory
1 Bungtown Road
P.O. Box 100
Cold Spring Harbor
New York, 11724 USA
Tel.: 1516 367 8455
Fax: 1516 367 8496
Central Email: pubaff@cshl.org
Central URL: http://www.cshl.org

Fels Institute for Cancer Research and
Molecular Biology
Temple University School of Medicine, 3307 N.
Broad Street
19140 Philadelphia
Pennsylvania, USA
Tel.: 1215 707 4307
Fax: 1215 707 1454
Central Email: preddy@sgil.fels.temple.edu
Central URL: http://www.medschool.temple.edu/
Fels Institute for cancer Research

Fred Hutchinson Cancer Research Center
1100 Fairview Avenue North
98109-1024 Seattle
Washington, USA
Tel.: 1206 667 5000
Fax: 1206 667 7005
Central Email: commrel@fhcrc.org
Central URL: http://www.fhcrc.org

H.Lee Moffitt Cancer Center & Research Institute
12902 Magnolia Drive
33612 Tampa
Florida, USA
Tel.: 1813 972 4673
Fax: 1813 972 8495
Central URL: http://www.moffitt.usf.edu

Ludwig Institute for Cancer Research—San Diego
University of California—San Diego
9500 Gilman Drive, Mail Code 0660
92093 La Jolla
California, USA
Tel.: 1858 534 7802
Fax: 1858 534 7750
Central URL: http://ludwig.ucsd.edu

Salk Institute Cancer Center
10010 North Torrey Pines Road
92037 La Jolla
California, USA
Tel.: 1858 453 4100
Fax: 1858 457 4765
Central Email: eckhart@salk.edu
Central URL: http://www.salk.edu

Shands Cancer Center at the University of Florida
1600 S.W. Archer Road
Box 100286 HSC
32610 Gainesville
Florida, USA
Tel.: 1352 395 05 58
Fax: 1352 395 05 72
Central Email: gail@surgery.ufl.edu
Central URL: http://www.ufl.edu/uf-active.html

Memorial Sloan-Kettering Cancer Center
1275 York Avenue
New York, New York, USA 10021
Tel.: 212 639 2000
Central URL: http://www.mskcc.org

University of Pittsburgh Cancer Institute
200 Lothrop Street
15261 Pittsburgh
Pennsylvania, USA
Tel.: 412 692 4670
Fax: 412 692 4665
Central Email: herbermanrb@msx.upmc.edu
Central URL: http://www.upci.upmc.edu

University of Texas M.D. Anderson Cancer Center
1515 Holcombe Blvd
77030 Houston
Texas, USA
Tel.: 1713 792 2121
Fax: 1713 799 2210
Central URL: http://www.mdanderson.org

UK

Cancer Research UK, National Office:
P.O. Box 123
Lincoln's Inn Fields
London WC2A 3PX, UK
Tel.: 020 7242 0200
Fax.: 020 7269 3100
Central URL: http://www.cancerresearchuk.org

Cancer Research UK Beatson Laboratories
Garscube Estate
Switchback Road
Bearsden
Glasgow G61 1BD
Scotland, UK
Tel: 0141 330 3953
Fax: 0141 942 6521
E-mail: beatson@gla.ac.uk

Christie Hospital NHS Trust
Wilmslow Road,
Manchester M20 4BX, UK
Tel.: 0845 226 3000
Central URL: http://www.christie.nhs.uk

The Ludwig Institute for Cancer Research
University College London Branch
91 Riding House Street, London W1W 7BS, UK
Tel.: 020 7878 4000
E-mail: webmaster@ludwig.ucl.ac.uk

Ludwig Institute for Cancer Research
Imperial College Faculty of Medicine
St Mary's Campus
Norfolk Place
London W2 1PG, UK
Tel.: 020 7724 5522
Fax: 020 7724 8586

GRAY Cancer Institute
P.O. Box 100
Mount Vernon Hospital
Northwood
Middlesex HA6 2JR, UK
Tel.: 01923 828611
Fax: 01923 835210
http://www.gci.ac.uk/

The Institute of Cancer Research
123 Old Brompton Road
London SW7 3RP, UK
Tel: 020 7352 8133
Fax: 020 7370 5261
Central URL: http://www.icr.ac.uk

  Chester Beatty Labs
  Fulham Road
  London SW3 6JB, UK

  Sutton
  15 Cotswold Rd, Belmont
  Sutton, UK Surrey SM2 5NG

The Medical Research Council
20 Park Crescent
London W1B 1AL
Tel.: 020 7636 5422
Central URL: http://www.mrc.ac.uk

The Weatherall Institute of
Molecular Medicine
University of Oxford
John Radcliffe Hospital
Headington
Oxford OX3 9DS, UK
Tel: 01865 222443
Fax: 01865 222737
Central URL: http://www.imm.ox.ac.uk

COMPANIES

AstraZeneca UK Ltd.
Horizon Place 600 Capability Green
Luton, Beds LU1 3LU UK
Tel.: 01582/836000
Fax: 01582/838003
email: medical.informationgb@astrazeneca.com
http://www.astrazeneca.co.uk

Pfizer Limited
Walton Oaks, Dorking Road
Tadworth, Surrey KT20 7NS UK
Tel.: 01737 331 111
Fax: 01737 332 507
http://www.pfizer.co.uk

Novartis Pharmaceuticals UK Ltd
Frimley Business Park
Frimley, Camberley
Surrey GU16 5SG,
UK
Tel.: +44/1276/698370
Fax: +44/1276/698449
http://www.uk.novartis.com

Schering-Plough Ltd
Shire Park
Welwyn Garden City
Herts AL7 1TW,
UK
Tel.: 01707 363 636
Fax: 01707 363 763
http://www.schering-plough.com

GlaxoSmithKline UK
Stockley Park West
Uxbridge, Middx UB11 1BT, UK
Tel.: 020 8990 9000
Fax: 020 8990 4321
email: customercontactuk@gsk.com
http://www.gsk.com

Procter & Gamble Pharmaceuticals UK Ltd
Lovett House, Lovett Rd
Staines, Middx TW18 3AZ, UK
Tel.: 01784 495 000
Fax: 01784 495 253
http://www.pgpharma.com

Genzyme Ltd
Haverhill Office, 37 Hollands Road
Suffolk CB9 8PU, UK
Tel.: 01440 703522
Fax: 01440 707783
http://www.genzyme.com

Chiron UK Ltd
PathoGenesis House, Park Lane
Cranford, Hounslow TW5 9RR, UK
Tel.: 020 858 04000
Fax: 020 858 04001
http://www.chiron.com

Genentech, Inc.
1 DNA Way
South San Francisco, CA 94080-4990, USA
Tel: 650 225 1000
Fax: 650 225 6000

ImClone Systems Incorporated
180 Varick Street
New York, NY 10014, USA
Tel.: 212 645 1405
Fax: 212 645 2054

Maxim Pharmaceuticals
8899 University Center Lane
Suite 400
San Diego, CA 92122, USA
Tel: 858 453 4040
Fax: 858 453 5005

GLOSSARY

Aflatoxin a carcinogenic compound produced by the mould *Aspergillus flavus* that contaminates some food products such as peanuts.

Alkylating agent a chemical that introduces an alkyl group onto DNA; they act as carcinogens but are also used in chemotherapy.

Allele an alternative form of a gene at the same locus or relative position in a chromosomal pair. One allele may be dominant over the other.

Angiogenesis the process of forming new blood vessels from pre-existing blood vessels by the growth and migration of endothelial cells in a process called "sprouting". The induction of angiogenesis is a hallmark of cancer.

Anoikis apoptosis triggered in response to lack of extracellular matrix ligand binding.

Antibody a protein produced by lymphocytes in response to an antigen and can specifically bind the antigen as part of an immune response.

Antigen a molecule capable of generating an immune response.

Antimetabolite an agent that resembles an endogenous metabolite and blocks a metabolic pathway.

Antioxidant a compound that significantly inhibits or delays the damaging action of reactive oxygen species, often by being oxidized themselves.

Antisense oligonucleotide synthetic nucleotide fragments that hybridize to complementary DNA or RNA in order to inhibit gene expression.

Apoptosis a process of "neat" programmed cell death. It plays a role in tumor suppression; inhibition of apoptosis is a hallmark of cancer.

Attenuated reduced virulence (infectivity) of a pathogenic micro-organism.

Autoimmunity a condition in which an individual's immune system starts reacting against one's own tissues, causing disease.

Basement membrane an acellular support of endothelial, epithelial, and some mesenchymal cells made up of a complex mix of extracellular matrix proteins, including laminins, collagens, and proteoglycans. It acts as a passive barrier that separates tissue compartments.

Benign characteristic of a tumor that does not invade surrounding tissues or metastasize.

Bioinformatics the use of computers and information technology to store and analyze nucleotide and amino acid sequences and related information.

Cachexia a metabolic defect often associated with cancer that is characterized by progressive weight loss due to the deletion of adipose tissue and skeletal muscle.

Cancer stem cells cells within a tumor that have the ability to self renew and to give rise to phenotypically diverse cancer cells.

Carcinogen a chemical or form of energy that causes cancer.

Carcinogenesis the process of inducing cancer.

Carcinoma a malignant tumor of epithelium.

cDNA the DNA sequence that is complementary to an mRNA.

Cell cycle the sequence of stages that a cell passes through between one cell division and the next. The cell cycle can be divided into four main stages: the M phase, when nuclear and cytoplasmic division occurs; the G1 phase; the S phase in which DNA replication occurs; and the G2 phase.

Chromatin fibers made up of DNA, RNA and protein that form chromosomes.

Chromosome a structure composed of a DNA molecule and associated RNA and protein. Humans contain 46 chromosomes in the nucleus of somatic cells.

Clonal originating from one cell.

Cytostatic a drug that stops cell growth.

Cytotoxic a drug that kills cells.

Differentiation the functional specialization of a cell as a result of the expression of a specific set of genes.

DNA response elements short sequences of DNA that act as binding sites for transcription factors in gene promoters.

Dominant negative a mutation that produces a protein that interacts with and/or interferes with the function of a wild-type protein.

Downstream refers to DNA sequences that are more 3′ to a point of reference. Note that by convention a DNA sequence is read from the 5′-end to the 3′-end.

Electromagnetic radiation a naturally occurring energy that moves as waves resulting from the acceleration of electric charge and the associated electric and magnetic fields. The characteristics of the radiation depend on its wavelength.

Electromagnetic spectrum the range of wavelengths over which electromagnetic radiation extends. The longest waves ($10^5 - 10^{-3}$ ms) are radiowaves and the shortest are gamma rays ($10^{-11} - 10^{-14}$ m).

Electrophilic molecules that are electron-deficient and therefore are attracted to compounds with a net negative charge.

Embryonic stem cells cells derived from the inner cell mass of an early embryo. When transferred into another early embryo they combine with host inner cell mass cells and contribute to embryo formation.

Epigenetic refers to inheritable information that is encoded by modifications of the genome and chromatin components that affects gene expression. It does not include changes in the base sequence of DNA.

Estrogens steroid hormones secreted by the ovary, but also produced by adipose cells, that act maintain female characteristics and act as a mitogen for breast cells.

Extravasation the process whereby a cancer cell exits a blood vessel or lymphatic vessel.

First pass organ the first organ en route via the bloodstream that lies downstream from the primary tumor site.

Gene a region of DNA that occupies a specific position on a chromosome and includes the regulatory region and coding region for a protein.

Gene amplification the multiple replication of a section of DNA that results in the production of many copies of the genes involved.

Gene expression the process by which the information encoded by a gene is converted for the making of a protein. In terms of molecular biology, this usually refers to transcription.

Genomics the study of all the genes contained in a set of chromosomes.

Genotoxic the ability of a substance to damage DNA.

Genotype the genetic characteristics of a cell or organism. Also the combination of the alleles at a particular locus.

Germline mutation a mutation in either egg or sperm cell DNA (as opposed to a somatic mutation). Mutations in germ cells only, can be passed on to the next generation.

Hematopoietic refers to tissue that can give rise to blood cells in the process of hemopoiesis.

Heterodimer a functional protein that is made up of two different subunits.

Heterozygous having different alleles at a given locus on homologous chromosomes.

Histones basic proteins within chromatin that bind DNA at regular intervals.

Homodimer a functional protein that is made up of two identical subunits.

Homozygous having the same two alleles at a given locus on homologous chromosomes.

Hypoxia a state of low levels of oxygen.

Incidence the number of new cases of cancers in a defined population over a defined period of time.

Intravasation the process whereby a cancer cell enters a blood vessel or lymphatic vessel.

Invasion spread to tumor cells into surrounding tissue.

Kinase an enzyme that transfers phosphate groups to a protein at serine, threonine, or tyrosine amino acids.

Knockout mice mice in which both alleles of a gene have been inactivated experimentally. These mice are often used to study gene function.

Lead compound a compound identified during the development of a drug that shows a desired activity, e.g. kinase inhibition.

Leucine zipper a protein domain that mediates dimer formation and is normally adjacent to a basic DNA-binding domain. It is characterized by a pattern of five leucine residues each separated by six residues.

Leukemia a type of cancer characterized by the overproduction of white blood cells or their precursors in the blood or bone marrow.

Ligand an agent that binds to a receptor. A specific hormone is a ligand for its corresponding hormone receptor.

Linear energy transfer (LET) rate of energy loss to the surrounding medium, in a radiation track (unit: keV/μm).

Lymphoma a solid tumor of T or B lymphocytes in the lymph nodes, thymus, or spleen.

Loss of heterozygosity loss of the second allele of a gene.

Malignant characteristic of a tumor that is capable of invading surrounding tissue and of metastasizing to secondary locations.

MAP kinases mitogen-activated enzymes that phosphorylate serine and threonine residues on proteins. Also known as extracellular signal-related kinases (ERKs).

Metastasis the process of cancer cells spreading from a primary site to secondary sites in the body.

Missense mutation a type of mutation that converts one codon to another, specifying a different amino acid.

Mitogen a substance that can cause cells to divide (i.e. undergo mitosis).

Mitosis the phase of the cell cycle whereby the cell divides to produce two daughter cells.

Morphology the study of form and structure of organisms.

Mutagen a chemical or form of energy that can cause a mutation.

Mutation a heritable change in the bases of DNA, which may include transitions, transversions, deletions, insertions, or translocations.

Necrosis a type of cell death characterized by membrane disruption and release of lytic enzymes. This "sloppy" way of dying contrasts cell death by apoptosis.

Non-genotoxic carcinogen a substance that causes cancer without damaging DNA.

Nonsense mutation a type of mutation that converts a codon that specifies an amino acid to one of the "stop" codons, thus signaling termination of translation and the formation of an incomplete polypeptide.

Nude mice immunodeficient mice (usually hairless) that have no cell-mediated immunity due to the absence of the thymus gland. They can be used experimentally to grow human tumors.

Oncogene a gene whose product is capable of transforming a normal cell into a cancer cell. Oncogenes result from the mutation of normal genes (proto-oncogenes).

Ontogeny the development of an individual.

Phagocytosis The process whereby particles or cells are engulfed by cells, such as macrophages. Cells that undergo apoptosis are consumed by phagocytosis.

Phenotype the observable characteristics of a cell or organism.

Phosphorylation the addition of a phosphate group (PO_4^{3-}) to a bio-molecule. Phosphorylation may cause conformational changes in proteins or activate particular enzymes.

Polymorphism the occurrence of two or more alleles for a given locus in a population where at least two alleles appear with frequencies of more than 1%.

Polyp a tumor that project from the surface of epithelial (e.g. polyps of the colon).

Prognosis a forecast or future outlook for a disease.

Promoter the regulatory region of a gene that initiates transcription; usually DNA sequences located 5′ to the coding sequences but may be located in other regions such as introns and 3′ sequences.

Protease an enzyme that degrades proteins.

Proteolysis enzymatic protein degradation involving cleavage of peptide bonds.

Proto-oncogene a normal cellular counterpart of a mutated gene that can cause tumors.

Purine the nitrogenous bases, adenine and guanine, found in DNA.

Pyrimidine the nitrogenous bases, cytosine, thymine, and uracil, found in DNA or RNA.

Radiolysis the use of ionizing radiation to produce chemical reactions.

Reactive Oxygen Species used to classify reactive intermediates of oxygen in this book (e.g. hydroxyl radicals, hydrogen peroxide, and superoxide radical) although broader definitions exist.

Receptor a transmembrane, cytoplasmic, or nuclear molecule that binds to a specific factor, such as a growth factor or hormone.

Recessive an allele that is expressed only when present in the homozygous or hemizygous state (i.e. two such alleles must be present).

Relapse reappearance of a disease.

Remission reduction in the severity of cancer as a result of treatment.

Retinoblastoma cancer of the retinal cells of the eye. A germline mutation in the retinoblastoma (*Rb*) gene is found in familial cases.

Sarcoma a malignant tumor of the mesenchyme, e.g. bone cancer.

Signal transduction the transfer of information along a pathway of a cell that converts a signal received from the outside of the cell to the inside, to generate a cell response.

Somatic cell All cells other than egg or sperm cells. Mutations in somatic cells cannot be passed on to the next generation.

S phase the phase of the cell cycle in which DNA synthesis occurs.

Self-renewal the process whereby a stem cell (or progenitor cell) gives rise to a daughter cell with equivalent developmental potential. For example, a stem cell divides to give rise to two daughter cells: another stem cell and perhaps another more differentiated cell.

Stem cell a cell that can self-renew and give rise to more differentiated cell types.

Supplements extra sources of dietary components taken in addition to food.

Telomere repeated DNA sequences that are located at the ends of chromosomes. The structures shorten upon each round of cell replication.

Telomerase an enzyme that extends telomere length. Elevated levels are observed in many cancer cells.

Therapeutic Index the difference between the minimum effective dose and maximum tolerated dose of a drug. The larger the value is, the safer the drug.

Transcription the process of transferring the information encoded by DNA into RNA; also referred the process that occurs when a gene is expressed.

Transfection the transfer of exogenous DNA into cells by experimental procedures such as microinjection or electroporation.

Transformation the changes that occur as a normal cell converts into a cancer cell.

Transgenic mice mice that carry foreign DNA, experimentally introduced, in every cell of its body.

Transition a DNA mutation whereby a purine (A or G) is exchanged for another purine (G or A) or a pyrimidine (C or T) is exchanged for another pyrimidine (T or C).

Translation the process of transferring the information encoded by RNA into protein using the genetic code.

Translocation a DNA mutation whereby the part of one chromosome is transferred to, or exchanged for, another part of a different chromosome.

Transversion a mutation whereby a purine is exchanged for pyrimidine or vice versa.

Tumor an abnormal growth of cells that can be either benign or malignant.

Tumor Suppressor Gene a gene whose product performs functions that inhibits tumor formation and therefore loss or mutation of (usually both copies of) these genes leads to tumor formation; also, a gene in which a germline mutation predisposes individuals to cancer.

Upstream refers to DNA sequences that are more 5′ to a point of reference. Note that by convention, a DNA sequence is read from the 5′-end to the 3′-end.

Wavelength a characteristic of a wave. It is the distance in meters between successive points of equal phase in a wave. For example: the distance between successive peaks.

Xenobiotics foreign substances to living systems.

■ INDEX